VOLUME 46 AUGUST 1985 PAGES 137–234

THE NATURAL HISTORY OF A SUMAC TREE, WITH AN EMPHASIS ON THE ENTOMOFAUNA

BY

DAVID G. FURTH
Division of Entomology, Peabody Museum of Natural History
Yale University
New Haven, Connecticut 06511, U.S.A.

TRANSACTIONS
The Connecticut Academy of Arts and Sciences

Published for The Academy by

ARCHON BOOKS HAMDEN, CONNECTICUT

First published for The Academy by
Archon Books, The Shoe String Press, Inc.
Hamden, Connecticut 06514

ISBN 0–208–02094–2

Printed in the United States of America

ACKNOWLEDGMENTS

I would like to thank the following specialists for identification of *R. tripartita* arthropods: U. Gerson (soil mites), E. Meshaloff (aphids), T. Rivnay (Chalcidoidea), D. Rosen (Chalcidoidea)—Department of Entomology, Faculty of Agriculture of the Hebrew University of Jerusalem (Rehovot); P. Amitai (Arachnida), G. Levi (Arachnida), H. Meinis (Mollusca), Y. Werner (Reptilia)—Department of Zoology, Hebrew University of Jerusalem (Jerusalem); Y. Ben-Dov (Coccidoidea), J. Halperin (Coleoptera, Isoptera), M. Sternlicht (Acarina)—Volcani Institute; A. Freidberg (Diptera), J. Kugler (Diptera, Hymenoptera)—Department of Zoology, Tel Aviv University; Z. Boucek (Chalcidoidea), A. Hayes (Macro-Lepidoptera), L. Mound (Thysanoptera), K. Sattler (Micro-Lepidoptera)—Department of Entomology, British Museum (Natural History); R. Linnavuori (Hemiptera: Heteroptera and Homoptera) (Finland). U. Gerson (Acari, Coccidoidea), Y. Ben-Dov (Coccidoidea), M. Sternilicht (Eriophyidae), D. Schweitzer (Lepidoptera, part), J. A. Slater and C. W. Schaefer (Hemiptera: Heteroptera) also checked the Appendix and provided useful information.

The following Israeli botanists provided useful discussions about *R. tripartita* and its populations: R. Dulberger (pollen measurement), J. Galil, and L. Nussbaum (Tel Aviv University); A. Danin, D. Heller, A. Shmida, and M. Zohary (Hebrew University of Jerusalem).

The visits to the Negev Desert populations were partially funded by the 1980–81 Teats Memorial Stipend from the Nature Reserves Authority of Israel. This research was partially conducted as a Hebrew University of Jerusalem Postdoctoral Fellow at the Department of Entomology, Faculty of Agriculture, in Rehovot.

ABSTRACT

Many aspects of the natural history of a Mediterranean/Saharo-Arabian Sumac species (*Rhus tripartita*), with ancient relationships to some tropical African species, were studied as a functioning ecological system. Detailed botanical and especially entomological observations, both in the field and laboratory over four seasons, revealed new specific and general Sumac relationships and the fact that this tree species exists in relict desert and refuge Mediterranean populations. There is a 2–6 week difference in phenology of the trees and their herbivores in these two distinct habitats. The herbivores (especially monophages and polyphages) and other tree-associates found on the roots, trunk, branches/twigs, leaves, fruits, and flowers are listed along with information on some of their reared parasites, predators, phenology, and other aspects of their biology.

Only relictual populations are under severe desert grazing pressure from domestic animals (goats and sheep) which makes these populations prime candidates for conservation programs. These multifaceted studies of *R. tripartita* elucidate the historical biogeography of this Sumac and some of its more host-specific herbivores and their coevolution.

An appendix includes a world-wide list from the literature of all arthropod herbivores and some other associates recorded from *Rhus, Pistacia,* and *Cotinus,* including hosts and geographical locations.

CONTENTS

LIST OF TABLES

LIST OF FIGURES

INTRODUCTION

This study was initiated while studying the biology and ecology of a flea beetle, *Blepharida sacra* (Weise) (Chrysomelidae: Alticinae), restricted to the Sumac *Rhus tripartita* (Bernard da Ucria) Grande. The author became interested in the other entomofauna associated with this tree and ultimately in all aspects of the ecology and natural history of this tree species. Several singular aspects of the entomofauna of *R. tripartita* have already been published (Furth and Halperin 1980, Furth 1982, Furth, Ben-Dov, and Gerson 1984). The present paper surveys this Sumac as an ecological system with a variety of associations, mostly entomological and some botanical, based on field studies carried out by the author in Israel and the Sinai between 1977 and 1981. Because this type of comprehensive study has not to my knowledge been done for any other member of this worldwide genus (*Rhus* Linnaeus) or its near relatives (i.e., *Cotinus,* Miller, *Pistacia* Linnaeus, *Schinus* Linnaeus), this study is intended to be a model for study of other members of these groups indicating potential types of ecological associations or phenomena. A cursory search of the literature for the above-mentioned four genera demonstrates how little is known about their entomofauna (Appendix), with the possible exception of the Pistacio Nut tree (*Pistacia vera* Linnaeus). Some of the fascinating history and folklore concerning Sumac (*Rhus* and *Cotinus*) has been included as part of the overall approach to the study of the genus.

Post (1932), and others as indicated, give the following synonymy for *Rhus tripartita* (Bernard. da Ucria) Grande: *R. oxycantholdes* Dum.-Cours., *R. oxycantha* Schousb. ex Cav. (Andrews 1952), *R. albida* Schousb., *R. syriaca* Boissier, *R. doica* Brouss, in Willd. (Andrews 1952).

The general distribution of *Rhus tripartita* is: Morocco (west, southwest near Atlantic coast) (Negre 1962, Quézel 1979); Algeria (mountains, west, meridonal Sahara, near Blidda coast) (Quézel and Santa 1963); Tunisia (central, south—Sahara zone only) (LeHouerou 1959); Canary Islands (Teneriffe) (Mouterde 1970); W. Sicily and Linosa Island (Diels 1898); Egypt (western

Mediterranean coastal area, northern Arabian (Eastern) Desert, Red Sea coast, Gebel Elba—southeast, south Sinai) (Tackholm 1974); Sinai Peninsula (central, north mountains) (Zohary, Heyn, and Heller, 1983); Sudan (Red Sea mountains) (Andrews 1952); Saudi Arabia (northeast—Gulf of Aqaba, northern and southern Red Sea coastal [hills]—Heja and Asir districts, Nejd district); Jordan (southern desert, Rift Valley area [hills and mountains]) (Zohary, Heyn, and Heller, 1983); Israel (Negev and Judean Deserts, Dead Sea Area [hills], Lower Jordan Valley [hills], western Samaria mountains); Lebanon (southern coast [mountains]) (Mouterde 1970).

Most existing information about *R. tripartita* is botanical and concerns records of its geographical distribution across North Africa into the Middle East. Further details of its botany, distribution, and phytogeography will be discussed in the Results and Discussion sections. Post (1932) and Zohary (1972) give general botanical descriptions, distributions, and some habitat and seasonal information about the Middle East populations. Danin et al. (1970, 1975) gives further distributional details for Israeli populations in the Negev and Judean Deserts. Another botanist, Pollak (1983), simultaneously studied one of the same populations of *R. tripartita* as the author, and discussed a number of botanical aspects in some detail, especially the phenology and reproductive biology. During the course of the four years of research on this rather restricted Sumac species, I have studied its phenology and some botanical aspects as well as the ecological associations. This has provided insight into the probable phytogeography and evolution of this species along with information for research on other plants of the region, especially Anacardiaceae.

The genus *Rhus* Linnaeus (*sensu lato*) is one of the largest and most complex genera of the Anacardiaceae. In this sense it contains about 250 species worldwide in a variety of taxa often considered as separate genera or subgenera: *Actinocheita* Barkley, *Cotinus* Miller, *Lobadium* Rafinesque, *Malosma* Nuttall ex Abrams, *Melanococca* Blume, *Metopium* P. Browne, *Ochoternacea* Barkley *Rhus* Linnaeus, *Searsia* Barkley, *Thezera* (D. C.) K. Koch, and *Toxicodendron* Tournefort (Brizicky 1963, Barkley 1965, and Young 1979). There has been much disagreement and discussion on the taxonomic placement of these and some other generic names (Barkley 1937, 1965, Brizicky 1962, 1963, Gillis 1971, Young 1979). It seems apparent that these genera, and possibly others (e.g., *Schinus* Linnaeus), form a somewhat related group within the Anacardiaceae.

Most recently Young (1979) has considered *Rhus* to contain two subgenera (*Rhus* and *Lobadium*) based on heartwood flavonoids. Brizicky (1963), analyzing external and internal morphological characters as well as some chemical ones, concluded that there are six valid subgeneric taxa of *Rhus*: *Rhus, Lobadium, Toxicodendron, Malosma, Thezera* (=*Searsia*), *Melanococca.* Anacardiaceae is most developed in the tropics and subtropics of both Old and New Worlds, especially in the Malaysian region. *Rhus* (*sensu lato*), however, is primarily distributed in the warm-temperate regions of Eurasia, North America, Africa, and also extending into tropical and cold-temperate regions.

Folklore

Rhus (*sensu lato*) and some of its close relatives are of ethnobotanical interest because they have been used since at least ancient Greek times for a variety of cultural practices. The two best known species are *Rhus coriaria* Linnaeus and *R. cotinus* Linnaeus (=*Cotinus coggygria* (Scopoli)). Coriar's or Sicilian Sumac (*R. coriaria*) is a common Mediterranean shrub and early Greek writers such as Dioscorides, Pliny, and Theophrastus discuss its uses for tanning, pharmaceuticals, and cooking (Barkley and Barkley 1938). The ground leaf of *R. coriaria* was best known for its effective properties for tanning hides and for dyes (Willis 1973), although it was also used as a medicine. Pliny wrote that this species was used for healing bruises, wounds, upset stomachs, canker sores, and some other medical ailments; he also mentioned its value for seasoning meats (Barkley and Barkley 1938).

The Venetian Smoke Tree (*C. coggygria*), also present in the Mediterranean and further east, was less known in earlier writings than Coriar's Sumac, but was also used for tanning and coloring hides, as a source of colored dye for silk, wool, etc., and of aromatic oils from the leaves and flowers (Brizicky 1962). Actually, the ethnobotany of Coriar's Sumac and the Venetian Smoke Tree was mentioned in most botanical treatises from Greek and Roman times through the Medieval and Renaissance periods and, in fact, these two were the only species of Sumac mentioned before 1600 (Barkley 1937).

Riley (1874) commented that commercial importation of European Sumac (*R. coriaria*) into the USA for tanning and coloring was then increasing, e.g., 12 million pounds in 1869. He mentioned that American Sumac production was increasing, e.g., Virginia was producing 6,000 tons per year which sold for $60 per ton. The three common American Sumac species of apparent economic value for tanning were Stag Horn Sumac (*R. typhina* Tournefort), Smooth Sumac (*R. glabra* Linnaeus), and Dwarf Sumac (*R. copalina* Linnaeus). The waxy oil extract from the fruits of Smooth Sumac and other Asian species can be made into a commercial vegetable wax, but this may also give a rather strong, pungent odor (Brizicky 1962, 1963). Leaf galls on some Sumac species yield special tannins used to make ink (Brizicky 1962). In Asia several species of *Rhus* (*Toxicodendron*) have long been used to produce a high grade Chinese and Japanese lacquer (Barkley 1937). The Indians of southwestern North America used several species of Sumac in different ways. They ate the fruits (ripe or green) and made both cool and hot drinks from them. All parts of the plant were used for medicines as remedies for coughs, chest pains, pinworms, hemorrhaging, and various other ailments. These Indians also used Sumac to make bows and baskets, as well as a tobacco-like mix (Barkley 1937). Besides the above-mentioned historical/ethnobotanical values of Sumacs, many species are useful for: ornamentals and ground cover; indicators of pioneer revegetation of denuded areas, especially deciduous woodlands; wildlife management; woodworking; etc. (Barkley 1937). More detailed information pertaining to the ancient history and ethnobotany of *Rhus* and relatives can be found in Barkley (1937) and Barkley and Barkley (1938).

The etymology of *Rhus* is somewhat obscured. Pliny said that it is derived from the Greek word (ϱοῦς) for these trees (=*R. coriaria*) meaning red and based on the red color of the fruit. Pliny also insisted that there is no Latin counterpart to this Greek derivation but some other explanations have also been offered for this dilemma (Barkley and Barkley 1938). The common English name of Sumac (Sumach) is similar to and probably derived from the Arabic word "Sommaq" or "Simmak" and the similar Byzantine name (Barkley and Barkley 1938).

Botany

Rhus (*sensu lato*) contains both shrubby and arborescent species, the differentiation is sometimes variable and subjective. The habitat varies but Sumacs are usually found on calcareous soils.

The root system is generally quite shallow. There is a common tendency towards rhizome-vegetative reproduction and often populations consist of clumps or clones; however, individual plants are also common and are the rule for some species (Barkley 1937). The leaves are alternate, varying from simple to trifoliate to multifoliate, and from thin to thick or leathery. The flowers are small, polygamodioecious, pentamerous, and form particulate clusters (Barkley 1937). The fruit is a drupe usually about as wide as long and sometimes somewhat flattened. The fruit exo and mesocarp are united and easily removable from the hard endocarp. The fruits are commonly reddish but also sometimes orange or brownish; some relatives of Poison Ivy have green or yellow-white to tan-colored fruits (Brizicky 1963).

The genus, whether considered in a broad or narrow sense, contains species with a variety of ecological parameters and environmental preferences. Although sexual (seed) reproduction is probably not as prevalent as with many plants, apparently there is evidence in most groups (subgenera) for seed dispersal by animals, especially birds (Brizicky 1962).

It has long been known that *Rhus* (*sensu lato*) contains high amounts of tannins. Further investigations have elucidated and analyzed the various tannins in greater detail. Of course, the poisonous species, i.e., *R.* (*Toxicodendron*), have attracted the most attention in biochemical and medical research. Recently, Young (1979) has surveyed the heartwood flavonoids of all (subgenera) of *Rhus* (*sensu lato*) and concluded that *Rhus* (*sensu stricto*) contained only two subgenera, the others being valid genera. The leaf flavonoids of *R. tripartita* and of some African relatives, analyzed by D. Young (Furth and Young, in preparation), indicate and verify some interesting relationships among the Sumacs and some of their herbivores. No doubt continuing research into the phytochemistry of *Rhus* and its relatives, especially regarding the defensive (secondary) compounds, will reveal further botanical relationships (even at the species level) as well as insight into the coevolution of the Sumac herbivore fauna.

MATERIALS AND METHODS

This study of *R. tripartita* was conducted during the four seasons from 1977 to 1981. Two major populations were visited rather regularly (2–3 week intervals) from September/October through January, and more sporadically afterwards. One population is located approximately one km east of Moshav Yarhiv (32°09′N, 34°58′E) and represents the Mediterranean environment. The second main population is in the Judean Desert at Wadi Qilt approximately 1–2 km west of Ein Qilt (31°50′N, 35°22′E), representing the desert environment. The following additional populations in Israel were visited at least once: Wadi Mukallik/Nachal Ogg (Judean Desert); Jiftlik—20 km north (Lower Jordan Valley); En Gedi (Dead Sea Area); Ma'ale Tamar, Makhtesh HaQatan, Makhtesh HaGadol, and Nachal Mamshit (Central Negev Desert); and Sanniriya (western Samaria mountains). Also, the population in Wadi Amlacha (SW Sinai Mountains) was visited in late 1977. During each visit phenological, ecological, and entomological observations were recorded and samples of entomological and botanical interest were brought back to the laboratory in Rehovot for further observational study, rearing, etc. The Sumac insects and spiders were collected with a 15 inch diameter canvas-bag net by sweeping or beating the tree branches and twigs. Many were also hand collected or aspirated directly from the trees. All Lepidoptera and some Coleoptera larvae were transferred to small plastic cages with a screened top for rearing adults and parasites in the lab. Details of rearing *B. sacra* and *T. jordana* can be found in Furth (1982) and Furth and Halperin (1980). Samples of fruit were collected periodically to survey for torymid seed wasps and their parasites. A supply of branches with leaves was collected during each visit to the two major populations for feeding laboratory stocks of herbivores, excess leaves were moistened and stored in a refrigerator. Color photographs of various stages of all arthropod associates and the botanical aspects of this Sumac were taken in the field and lab. Adults of some species of the insects and spiders were taken or sent to specialists in Israel and Europe for identification. These specialists are listed in the Acknowledgments section. Insect voucher specimens are deposited at Tel Aviv University, Department of Zoology.

Several soil samples were collected at each of the two main populations from under the Sumac tree canopy and in open areas, then analyzed for soil type and organic matter content. Soil samples from W. Qilt (under the tree canopy and two meters away) were collected and run through a Berlese Funnel in the lab to survey for soil organisms, especially mites.

The list of arthropods feeding on *Rhus, Cotinus,* and *Pistacia* (Appendix) was prepared by searching the following: the Biological Abstracts (from the first volume); the selected bibliography of insect-vascular plant studies (Kingslover et al. 1984), random entomological literature at Yale University and the Connecticut Agricultural Experiment Station in New Haven, and many of the references with a host plant index listed in the similar literature survey of fern-feeding arthropods of the world (Balick, Furth, and Cooper-Driver 1978); and some specialists were also consulted. An attempt has been made to update the arthropod nomenclature and synonymy, wherever possible. It is considered impractical to include author names for all foodplants, many are included in the cited references. The geographical abbreviations are self-explanatory, those for states of the USA follow the standard two letter postal service abbreviations. Polyphagous arthropod species are indicated when known or when comprehensive literature on the genus, family, or order was readily available. In such a literature search, there is always a chance that some records are accidental, i.e., not actually consistent herbivores on the plant on which they were found; however, there are probably very few such cases in the Appendix list. This world list of arthropod herbivores of Sumac and relatives is not supposed to be complete, especially because so much herbivore-foodplant information is buried in taxonomic monographs and various surveys that do not contain food plant indices. However, this list contains the majority of the recorded *Rhus-Pistacia* feeders in the world and should provide a useful basis for study of the herbivores of these plants.

RESULTS

Botany

Rhus tripartita is a dioecious tree often with densely packed branches and twigs. The trees can reach a height of approximately 4 meters and may have broad lateral growth of twisted trunks and branches (Fig. 1). Most commonly it is known from populations exposed to heavy grazing by domestic animals, i.e., goats and sheep, and does not grow more than 1–1.5 meters high (Fig. 2). This species is one of the few Sumacs that possess "pseudo-thorns" formed from small twigs (Figs. 11 and 16) or ends of larger twigs [*R. pentaphylla* Linn. also apparently has such thorns to a lesser degree (Zohary 1972)]. The leaves are trifoliate, smooth-surfaced, the margins obtusely lobed and the leaf petiole is slightly shorter than the leaflets (Fig. 10). The leaves are up to 2 cm in length, sessile and obovate-cuneate in shape (Zohary 1972). The leaf color is dark green which gives the tree a very dark, rather distinctive appearance from a distance, making it easily detectable in desert environments. The flowers are small, light green or yellow-green, with short terminal racemes. The staminate inflorescences are often larger than the pistillate. The flowers have 5 petals and a 5-lobed calyx. The staminate flowers have 5 stamens with pollen-producing anthers and the pistillate flowers have 5 vestigial anthers; in both cases these stamens originate between the bases of the petals. The pollen is yellow-orange in color, and sticky. The average pollen size of 11 samples of a population from western Samaria is 17.5 microns wide and 33.8 microns long. All flowers have a central disk with floral nectaries. The female flower has 3 styles protruding from the bulbous ovary. The fruits are globular and smooth-surfaced, somewhat laterally compressed, red and sometimes partly green in color. New twig growth is rather soft and flexible and has a dark reddish-brown color as opposed to the hardened gray-colored older twigs and branches. Senescent leaves and fruit either on or off the trees are also a reddish-brown color, as are the trunk heartwood and the roots.

The desert or arid zone populations (i.e. Sinai, Negev, Judean Deserts, and Jordan Valley) live under similar environmental conditions of temperature and moisture regimes, and, to some degree, soil. These populations differ significantly in seasonality and phenology from the large, more or less continuous population in western Samaria. There are also different ecobotanical associations of *R. tripartita* in these two rather different habitats, as well as differences and similarities in the ecozoological associations, especially the entomofauna.

In the representative arid zone population studied in Wadi Qilt the trees grow on the often steep-sided slopes or sides of the canyons; never in the canyon bottom or on top of the mountains flanking the canyons (Fig. 4). In the Negev, populations often grow out from small fissures in hard limestone or dolomite rock faces. In the Sinai Mountains the large trees grow among the granite boulders of rather shallow canyons that have an extremely steep gradient. Almost all the populations in the Sinai, Negev, and Judean Deserts are small (fewer than 10–20 individuals) and very isolated from the nearest population; W. Qilt is the exception.

The environment of Wadi Qilt is quite harsh, with an average annual temperature of 23–25°C, an average annual precipitation between 100–200 mm (Jericho has 144 mm), and an average annual potential evaporation of 1800 mm. Soil analysis of the W. Qilt population revealed a sandy textured, calcareous, brown desert skeletal soil found in shallow pockets. This soil normally contains less than 1% organic matter, but directly under the *R. tripartita* canopy may contain 3–5% organic matter, and thus is darker in color. To underscore this contrast in soil organic matter content, W. Qilt soil samples taken in winter from under a Sumac tree canopy and from 2 meters away were run through a Berlese funnel. The sample under the canopy yielded soil mites (Acarina) from 12 families and 4 orders, while 2 meters away only one species was recovered.

Danin et al. (1970), Zohary (1972), and others consider *R. tripartita* to be a phytoecological codominant with *Retama retama* (Forssk.) Webb. (Papilionaceae). However, the Judean Desert populations are usually dominant in the areas where they grow and no significant amount of *Retama* was observed among the large population in Wadi Qilt (possibly because of overgrazing). The climate and various ecological associations [botanical—certain species of annual Cruciferae and Boraginaceae; zoological—

Leaf Beetles (Furth 1979, 1980)] in Wadi Qilt indicate that it is a warm arid subtropical environment of the Saharo-Arabian/ Eremian zone.

The population of *R. tripartita* studied in western Samaria near Moshav Yarhiv, is basically Mediterranean and represents a quite different environment from W. Qilt. The average annual temperature is approximately 21°C, mean annual precipitation 557 mm (Horeshim) (Israel Meteorological Service 1967), and the annual evaporation potential is approximately 1100 mm (Ashbel 1973). The soil at the Yarhiv site consists of a calcareous terra rosa formed on a Cenomanian limestone with a clay texture. The upper 5 cm of soil at this site normally contains 3–5% organic matter, but under the canopy of *R. tripartita* it may reach 10–12% with a darker brown color. This is among the highest organic matter levels for soils measured in Israel.

The area around the Yarhiv site is a Mediterranean garigue (batha or phrygana) with low and patchy populations of mostly shrubby (often spinous) vegetation. Much of the area has been disturbed by human settlements, agriculture, and reforestation. Despite this, the region supports the largest continuous stand of *R. tripartita* in Israel, stretching from around Qalqilya in the north about 15 km to south of Rosh Ha'ayn. The population is about 11 km wide east to west and located east of the Azzun-Lubban Road, and is very dense around Sanniriya (Fig. 5) and Nahal Kana, where it ascends the Samaria mountains. In the flatter areas near Yarhiv, the trees grow on rocky hills near the coastal plain and on the gentle slopes of the Samaria foothills to the east (fig. 6). This is somewhat in contrast to the steep-sided rocky slopes of the desert populations.

The phytoassociations of these Mediterranean garigue populations vary. There is little *Pistacia lentiscus* L. or *Ceratonia siliqua* L., common in many Mediterranean garigue areas. *Rhamnus palaestinus* Boisseri, *Capparis spinosa* L., and several other shrubs occur, but *R. tripartita* is always the dominant plant species in these areas. Another interesting difference between the arid environment populations and these Mediterranean garigue populations is that the average tree height of the latter Sumac populations is approximately 2 meters.

The seasonality and phenology of *R. tripartita* varies somewhat according to the habitat and its environmental conditions, i.e., Mediterranean or desert. Four seasons of monitoring aspects of the botany and herbivorous insects of the populations at W. Qilt and Yarhiv made it apparent that there are significant phenological differences. In the arid environment of Wadi Qilt the Sumac trees are: deciduous from late June to early October; producing new leaves in early October (rarely late September), with most trees reaching full foliage between late October and mid-November; flowering from mid-November into January, with the peak season during most of December; fruiting in the last half of December through February; producing a noticeable amount of new twig growth from January to April. In the Mediterranean environment: this Sumac is not truly deciduous because, even though some trees lose some leaves, most have green leaves and are even producing some new leaves in the hot dry summer months, however, the main surge of new leaf growth is from late September through November; new twig growth is especially prominent in April and apparently begins in January or February; flowering is from early October through January, but is most prevalent from late October through November and some of December; fruits develop from the later half of October through January, especially during November and December.

Apparently the relatively more mesic environmental conditions in the Mediterranean garigue enable the Yarhiv population to have an earlier, more progressive, and longer season than the Judean Desert populations. The foliage begins and fills out approximately two weeks earlier in the Mediterranean population. Flowering in the Mediterranean environment is about four weeks in advance of the desert populations and fruiting progresses even further at Yarhiv than at W. Qilt within the first six weeks; thus, a type of cumulative effect of two weeks for each major botanical stage (i.e., leaves, flowers, and fruits). There is, of course, some overlap in these life stages of the Mediterranean and desert populations, but there is a distinctly increasing advancement in the seasonality and phenology of the Mediterranean population. This advanced phenology is also apparent in the life cycles of certain herbivorous insects associated with *R. tripartita* (e.g., *Blepharida sacra*), but it is more difficult to quantify. The 1979–1980 season was a representative season and Tables 1–3 demonstrate graphically the detailed observations of botanical phenology from the desert (W. Qilt) and Mediterranean (Yarhiv) sites. These types of observations were recorded as often as possible, each year for four seasons.

Table 1. *Rhus tripartita*: Comparative Botanical Phenology—Leaves (1979–80), per average tree on observation date

+ = Yarhiv Population (= Mediterranean Refugia)
● = Wadi Qilt Population (= Desert Relict)

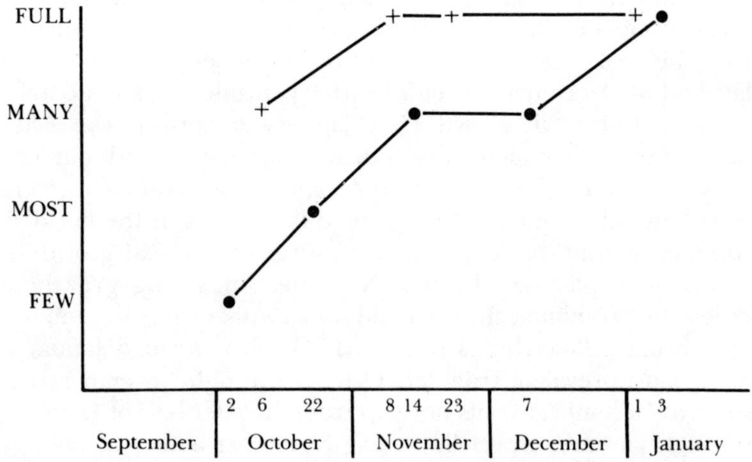

Table 2. *Rhus tripartita*: Comparative Botanical Phenology—Flowers (1979–80), per average tree on observation date

+ = Yarhiv Population (= Mediterranean Refugia)
● = Wadi Qilt Population (= Desert Relict)

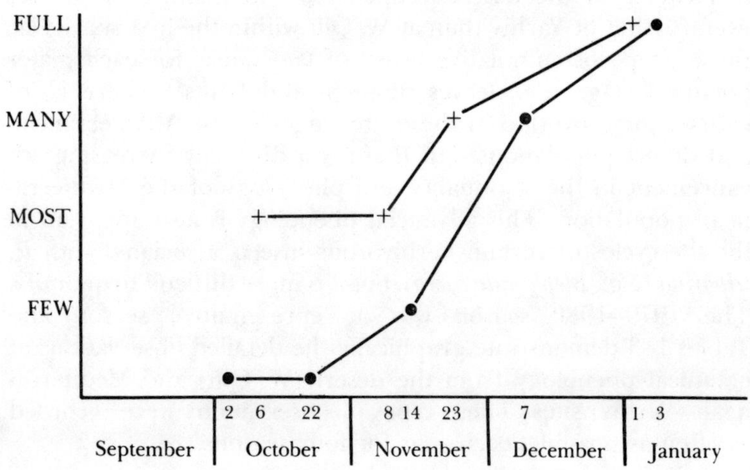

Table 3. *Rhus tripartita*: Comparative Botanical Phenology—Fruits (1979–80), per average tree on observation date

+ = Yarhiv Population (= Mediterranean Refugia)

● = Wadi Qilt Population (= Desert Relict)

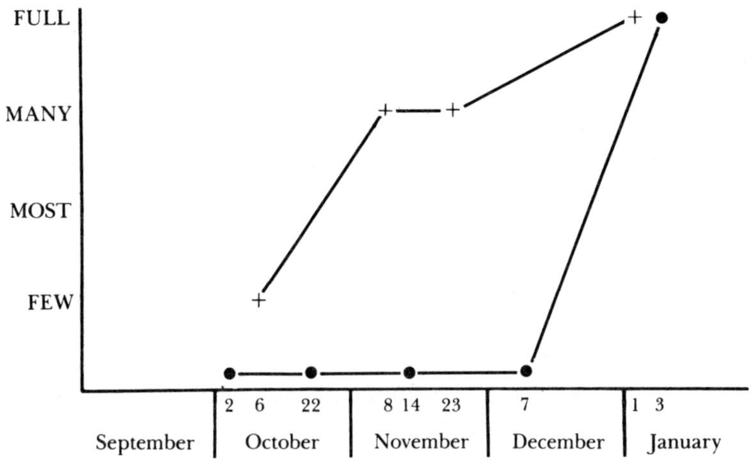

ENTOMOFAUNA

The major emphasis of this research on *R. tripartita* was to survey the entomofauna directly and indirectly associated with this Sumac. Several species of the Sumac insect fauna were collected and reared during each of the four seasons (some yielding parasite, life history, and other biological information), others were found less regularly. Some insects were host specific to *R. tripartita,* others apparently oliphagous on other Anacardiaceae (e.g., *Pistacia*), and still others apparently polyphagous. The level of food plant preference or specificity is not always definitively known at this point and may require more extensive field survey and literature research. However, it is believed that the arthropods listed in Tables 4–7 represent the major part of the *R. tripartita* system. This study attempts to include as many aspects of the ecology of *R. tripartita* as possible, i.e., botanical, zoological, edaphic, climatological, etc. Both natural and unnatural population pressures are also considered.

From the Tables (4–7) it is evident that most Sumac arthropods are restricted to either the desert or Mediterranean population; however, a few occur in both environments (e.g., *Blepharida sacra, Megastigmus pistaciae, Eutelia adulatrix, Ophiusa tirhaca, Lygus conspurcatus*). In other cases a particular niche is filled by different members of the same group (family), e.g., leaf-feeding gelechiid leaf-rollers and cicadellid leaf hoppers, flower thrips, crab spiders, nectar-feeding ants, and micro-Diptera/Hymenoptera pollinators.

The associations (both specific insect herbivores and casual associates) with the various plant parts of *R. tripartita* are as follows:

Roots and Trunk (Table 4)

On January 3, 1980 *Camponotus thoracicus fellah* (Hymenoptera: Formicidae) were seen scurrying about and entering the soil near some Sumac plants in W. Qilt. Upon investigation, it was discovered that they were tending mealybugs on the roots of *R. tripartita.* On December 18, 1980 this mealybug was found again without attendant ants. Egg masses, immature crawlers and adult females of the mealybug were found feeding on the roots and underground part of the crown of the plants. After considerable study of this mealybug by Y. Ben-Dov and U. Gerson, it was determined to be a new species, *Peliococcus deserticola* (Pseudococcidae) (Furth, Ben-Dov, and Gerson 1984) (Fig. 7). This mealybug was also found earlier in the season (22 October) at W. Qilt on the leaves of *Onosma aleppica* (Boraginaceae), possibly an alternate host. Root feeding by this mealybug on *R. tripartita* is probably an obligate stage of its life cycle and at least part of the reason that it was undiscovered for so long, although it was possibly found by F. S. Bodenheimer in the 1920s on the roots of *R. tripartita* in a different location in the Judean Desert (see Furth et al. 1984).

Also in December, in the Judean Desert, *Microcerotermes diversus* (Termitidae) swarms and settles into underground activity (nest building and foraging). In the region of W. Qilt many of these termites eat and tunnel through the root systems of *R. tripartita* and proceed up into the trunk and even along the lower branches of the tree. When these termites reach the thin branches they create a system of hollow tubes (apparently made of regurgitated soil and pulp) attached to the outside of the trunk or branch through which they are able to crawl and continue to feed on the branch. These mud-like tubes presumably also offer protection for these soft-bodied insects from the elements and predators. These termites are part of the desert—*R. tripartita* ecosystem but are not restricted to this plant species. In the Negev Desert Sumac populations there are trees with branches and trunks severely damaged by termites.

Another trunk pest is *Scobicia chevrieri* (Bostrichidae) found in W. Qilt (J. Halperin, personal communication) and in the Negev. The larvae feed primarily on dead wood of the trunks and large branches, creating characteristic small round exit holes in the wood. A few buprestid larvae, boring in live trees, were found in tree trunks at W. Qilt, possibly *Capnodis* sp.

Branches (Tables 4 and 7)

The branches and twigs of the W. Qilt population often have large infestations of the scale *Melanaspis inopinatus* (Diaspididae) from October through January, also recorded in Israel from *Pistacia lentiscus, P. atlantica,* Desf. *P. palaestina* Boiss., as well as Almond and *Styrax officinalis* L. (U. Gerson, personal communication). The males are narrow, white, and about 2 mm long and are sometimes found on the leaves early in the season (October). The females, more abundant later (late October till November), are round and larger (about 4 mm diameter). Females are black or dark gray and are sometimes in such large numbers that little of the branch surface is visible (Fig. 8). A species of parasitic wasp (Encyrtidae) and a predatory beetle (*Cybocephalus* sp., Nitidulidae) were found attacking this scale at the W. Qilt population.

The branches are the home for many of the general predators,

Table 4. Fauna of *Rhus tripartita*—Roots, Trunk, and Branches

Species	Plant Part/Parasite	Mediterranean	Desert
Melanaspis inopinatus Leonardi	branch	–	+
encyrtid sp.	parasite (*M.i.*)	–	+
Cybocephalus sp.	predator (*M.i.*)	–	+
Scobicia chevrieri (Villa)	trunk (weak/dead)	–	+
buprestid sp.	trunk	–	+
Microcerotermes diversus Silvestri	trunk/roots	–	+
Peliococcus deserticola Ben Dov and Gerson	roots	–	+
Camponotus thoracicus fellah Emery	tending *Peliococcus*	–	+

– = absent
+ = present

especially the spiders, that are regularly associated in certain ways with *R. tripartita* at different seasons. In the desert populations of W. Qilt *Stegodyphus* sp. (Eresidae) forms long tube-like webs along the length of the branches with carcasses of its prey attached to the sides. This spider is active from late September through November and possibly longer. *Steatoda maura* (Simon) (Therididae) makes a more open web between branches and is common from late October through November and probably December. Another small spider (Dictynidae) is common in W. Qilt late in the season from April until leaf drop in June and makes its web at the tips of twigs or small branches. Also, on the branches of the larger trees at W. Qilt each year there were egg masses of the neuropteran spider egg parasites—Mantispidae; however, in each case the eggs had already hatched. In April an adult of *Mantispa* sp. was found in a tree that had regularly produced mantispid egg masses. In the Mediterranean population at Yarhiv only *Eusparssus* sp. (Sparassidae) occurred repeatedly in October; forming long, tapered, funnel-like webs. The above-mentioned spiders are regular members of the *R. tripartita* community even though all can also be found in other associations. There are other spiders in Mediterranean and desert populations that were occasionally collected but with no regularity. In the autumn, larger branches of the W. Qilt population usually had egg masses of *Mantis* sp., a general predator in the region, a full-grown adult was found in April.

The semiparasitic perennial plant, *Loranthus acaciae* Zucc. (Loranthaceae), known primarily from species of *Acacia,* is well established on several of the larger Sumac trees at W. Qilt. It is rooted on the branches, its long red flowers are prominent from late November until January, and its oval sticky fruits are dispersed by birds that perch or nest on these larger trees. This parasitic plant is spreading in the W. Qilt Sumac population (Fig. 9).

The dense thorny branches of *R. tripartita* at both the W. Qilt and Yarhiv populations support nests of unidentified bird species; they are always located near the center of the tree and are well protected.

An interesting associate of Sumac in the Judean Desert population at W. Qilt is the Great Grey Shrike (*Lanius excubitor* Linn., Lanidae). This predatory bird is well known for using the thorns of trees, especially acacias, and even barbed-wire for killing and impaling its prey. In the winter months (October through January) some of these shrikes inhabit the warmer southern areas of Israel. In W. Qilt the most significant thorned objects are the branches and twigs of *R. tripartita* and the shrikes use the thorns of the larger trees (1–2 meters) to impale their prey, which are then quite visible (Fig. 16). The following animals have been found impaled on Sumac thorns at W. Qilt by the Great Grey Shrike, often in a fresh condition: the gecko, *Ptyodactylus hasselquistii guttatus* Heyden, the lizard, *Eremias guttulata* (Licht.), the frog, *Rana ridibunda Pall.*, the yellow scorpion, *Scorpio maurus palmatus* (Hemprich & Ehrenberg), the Oriental Hornet, *Vespa orientalis* Fab., the darkling beetle, *Adesmia* sp. (Tenebrionidae), and the snail, *Helix engaddensis* Bgt. It has been suggested by some that this shrike may also impale some of these animals as a way of display to their mates that they are able to provide enough food and/or as a method of marking their territory; the impaled prey are often not eaten and some consist of distasteful insects, e.g., *Meloe* sp. (Meloidae).

Leaves (Table 5)

As expected, leaves provide the biggest resource and support the largest diversity of the *R. tripartita* herbivorous fauna. This study was initiated with research on the biology and ecology of perhaps the most restricted of these herbivores, *Blepharida sacra* (Chrysomelidae) the Sacred Sumac Flea Beetle, originally described in 1897 from W. Qilt. The food plant of this rather large flea beetle was unknown until discovered by the author in 1974 in W. Qilt. This beetle has now been found in the Judean Desert populations (e.g., W. Qilt and Nahal Ogg) as well as in the western Samaria foothills (e.g., Yarhiv); it is absent from other smaller more isolated Judean Desert populations and from the Negev and Sinai Desert populations of *R. tripartita.* The adults are 6–8 mm long with a reddish brown background color (like the young twigs and senescent leaves of its Sumac host) with lengthwise thin yellow stripes on its upper side. The largest number of adults are found during the mating season (September–October) and after adult emergence (May–June), before the summar diapause (July–September). The green and white striped larvae can be found actively feeding on the leaves from October through December. The larvae appear moist, like slugs, and have a long "thread" of feces trailing from their posterior end (Fig. 10). There is a prepupal/pupal diapause in subterranean cells near the base of the Sumac trees, lasting up to 7.5 months, depending on environmental conditions. The adults lay packets of eggs on the branches covered by hardened fecal material and these packets can remain attached to the branches a year or more after hatching. An egg parasite (Eulophidae;: *Tetrastichus* sp. near *gallerucae*) and a larval parasitic fly (Tachinidae: *Meigenia mutabilis*) have been recovered from this beetle at both the Judean Desert and Samaria populations. The phenology and life cycle of the W. Qilt and Yarhiv populations of *B. sacra* differ: the Yarhiv poulation is usually two to three weeks more advanced in development. Further details of biology and ecology of this beetle are given elsewhere (Furth 1982).

Another foliage herbivore restricted to *R. tripartita* is the endemic processionary caterpillar *Thaumetopoea jordana* (Thaumetopoeidae) known only from the lower Jordan River Valley and adjacent arid regions of Israel and Jordan. Adults are active in October through November and are sexually dimorphic. Females lay a cluster of eggs (averaging 200 eggs each) covered by a layer of abdominal scales on a branch or trunk. The egg cluster may remain evident on the trees for several years after emergence. The larvae hatch, feed, and remain together as a brood throughout the season (Fig. 11). They form processionary groups defoliating sections of the Sumac trees and often causing severe local damage. The dense larval hairs are toxic (urticating) to the touch often causing a strong allergic reaction. Most larvae at W. Qilt are actively feeding from November through January. Similarly to *B. sacra, T. jordana* has a long summer diapause in the subterranean pupal stage, 6–8 months between February and September, some even diapause 14–18 months and emerge the following season. An egg parasite (Encyrtidae: *Ooencyrtus* sp. near *masii*) and a larval parasite (Tachinidae: *Palesisa nudioculate*) were recorded from this processionary caterpillar at W. Qilt. For further details on the biology of this moth see Furth and Halperin (1980).

A number of other less host specific leaf herbivores were collected (some reared) from either the Judean Desert or Mediterranean populations or both (Table 5). The microlepidopteran Gelechiidae are leaf rollers and tiers and are represented by a different species in the desert populations than in the Mediterranean populations. Apparently the youngest instars feed on just opened leaf sprouts; later instars (green, apically tapered, and very active—"jumping" and twisting violently when disturbed) tie or roll leaves together and feed on the enclosed leaf surface

(Fig. 12). Also, not uncommonly, they will tie the leaves to fruit or fruit clusters, then feed on fruits from the outside or even bore into the fruit and feed. The last instar is usually brownish or purplish in color. In the Mediterranean populations the larvae of *Stenovalva disclusa* are found from late October until early April with a pupal time of 14–39 days. This species probably breeds continuously with several generations each year. In the desert populations, *Teliopsis sarchochroma* larvae are found in October and from late February well into April, with a pupal time of 13–35 days. This species probably has at least two generations per year.

Another moth (Pyralidae: *Noorda* sp.) found only at Yarhiv has green larvae with subdorsal white lines, is somewhat longer than the gelechiid larvae and feeds primarily on clusters of ripening fruit, but may also feed to a small degree on leaves. These pyralid larvae are active from late October until December with a pupal period of 20–39 days and apparently only one generation per year. A parasitic fly (Tachinidae: *Peribaea tibialis*) and an unidentified species of braconid wasp were reared from larvae of *Noorda* sp.

The macro-lepidoptera, excluding *T. jordana,* feeding on leaves are represented by *Ophiusa tirhaca* (Noctuidae), a large *Catocala*-related moth with green and brown forewings and yellow underwings with a transverse black band. Its larvae are found feeding on this Sumac at both W. Qilt and Yarhiv. The earlier instars of this species are gray (similar to older twigs and branches) but the later instars are intricately patterned reddish-brown and are well-camouflaged on the newer twigs and branches. Larval development is approximately two months during late October through early January; the pupal period is 60–79 days. Another noctuid, *Eutelia adulatrix,* is smaller than the previous species with a mottled brown pattern and also found in both major environmental populations of the Sumac; at Yarhiv in October and November and at W. Qilt in April. The larvae are green with a pair of sub-mid-dorsal white lines, becoming almost deep purple in the last instar. The pupal period is 19–47 days for the Yarhiv population and the specimen collected in mid-April from W. Qilt evidently went through a long diapause and took 4.5 months to emerge in early September. The larvae of a geometrid, *Paracostis syntropha,* are found in late November at Yarhiv and have a pupal period of 29–31 days. A geometrid species with cryptic-colored larvae is found in the desert populations (Negev) and has a larval period of approximately 1.5 months; however, no moths emerged from pupae, thus identification of this species was not possible. A species of leaf mining Lepidoptera is present in the Mediterranean population in October and April; however, it remains unidentified because no moths were successfully reared.

Several Hemiptera are of varying importance as leaf feeders. The plant bug *Lygus conspurcatus* (Miridae) is found at Yarhiv (October till early January). It is extremely abundant and certainly a major pest of this Sumac at Yarhiv in October and November with nymphs also present then. It feeds on leaves, but quite often is found on flowers. Of the Homoptera, two species of Cicadellidae are important. The yellow and red patterned *Zygina discolor* is found at W. Qilt from November through April, most abundantly in January. A species of scale, *Ceroplastes* sp., is found on leaves at Yarhiv in October and November, often tended by ants (*Crematogaster* sp.). The polyphagous whitefly, *Acaudaleyrodes citri* (Aleyeurodidae), is occasionally on the leaves at both Yarhiv and W. Qilt in November and April and possibly throughout the season. Two polyphagous species of aphids, *Aphis gossypii* and *A. craccivora* (the former is much more abundant) are found feeding on the new twig emergent leaf sprouts in April at Yarhiv. The only other herbivorous beetle besides the Sumac Flea Beetle is the polyphagous weevil, *Lixus nubilosus,* the adults of which are found at Yarhiv from October through April, most abundantly in October and November. This weevil is well camouflaged on new twigs with its mottled red-brown color. A rather enigmatic mite, probably a species of *Aculus* (Acarina: Eriophyidae) creates leaf galls, blister-like eruptions, on Sumacs at Yarhiv (Fig. 13) and is found abundantly in October and January and also in June and July; this mite may have an alternative or intermediate host plant (M. Sternlicht, personal communication).

Fruit (Table 6)

As mentioned above, the gelechiid leaf-tiers also often feed on the fruit (*Teleiopsis sarcochroma*—deserts, and *Stenovalva disclusa*—Mediterranean). An even more significant fruit feeder that eats leaves occasionally is the pyralid moth, *Noorda* sp. However, the main pest of *R. tripartita* fruits is the oliphagous chalcidoid wasp *Megastigmus pistaciae* (Torymidae) which, as its name implies, also attacks *Pistacia.* Adult wasps are found in the Mediterranean populations from October till January and in the desert (Negev) in March. Infested fruits collected at Yarhiv in early and late

Table 5. Fauna of *Rhus tripartita* —Leaves

Species	Plant Part/Niche/Parasite	Mediterranean	Desert
MONOPHAGES:			
Blepharida sacra (Weise)	leaf	+	+
Tetrastichus gallerucae ? (Fonscolombe)	egg parasite (*B.s.*)	+	+
Meigenia mutabilis Fall.	larval parasite (*B.s.*)	+	+
Thaumetopoea jordana (Staudinger)	leaf	−	+
Ooencyrtus masii ? Mercet	egg parasite (*T.j.*)	−	+
Palesisa nudioculate Villeneuve	larval parasite (*T.j.*)	−	+
Teleiopsis sarcochroma Walsingham	leaf	−	+
Stenovalva disclusa Meyrick	leaf	+	−
OLIGOPHAGES:			
Eutelia adulatrix (Huebner)	leaf	+	+
Noorda sp.	leaf/fruit	+	−
Peribaea tibialis (Robineau-Desvoidy)	larval parasite (*Noorda*)	+	−
Aculus ? sp.	leaf	+	−
Paracotis syntropha (Prout)	leaf	+	−
Imbecilla imbecilla (Linnavuori)	leaf	+	−
microlepidoptera sp.	leaf (miner)	+	−
Ophiusa tirhaca (Cramer)	leaf	+	+
POLYPHAGES:			
Lixus nubilosus Boheman	leaf	+	−
Cryptocephalus fulvus Goeze (Chrysomelidae)	leaf	+	−
Lygus conspurcatus Reuter	leaf/flower	+	+
Zygina discolor (Horvath)	leaf	−	+
Aphis gossypii Glover	leaf	+	−
Aphis craccivora Koch	leaf	+	−
Acaudaleyrodes citri (Priesner and Hosny)	leaf	+	+
Ceroplastes sp.	leaf	+	−
Crematogaster sp.	tending *Ceroplastes* sp.	+	−

− = absent
\+ = present

November emerged in late December and early January; those collected in mid-May began emerging in late September until mid-October. Infested fruits appear discolored from their normal light green or partly red color and are entirely dull orange or purple. There is only one wasp per fruit; a sample of 128 fruits collected at Yarhiv in mid-May had 61% infestation (Fig. 14). A large percentage of fruits observed in a Negev Desert population on 13 March were also infested and emerged 21 March. At Yarhiv there are two other parasitic wasps associated with the fruits and *M. pistaciae: Mesopolobus* sp. (Pteromalidae) parasitizes larvae of *Megastigmus* and *Eupelmus urozonus* (Eupelmidae) parasitizes either the larvae of *Megastigmus* or is hyperparasitic on larvae of *Mesopolobus*—both of these parasites emerge over approximately the same period as *Megastigmus* (September/October and December/January).

Flowers (Tables 6 and 7)

A wide variety of insects are attracted to flowers of *R. tripartita* but few are found with any regularity feeding on the flowers. Because of the dioecious nature of this Sumac, there are probably certain insects attracted more to male flowers (e.g., pollen-feeders) and others attracted to female flowers. Both male and female flowers have floral nectaries that attract insects, especially ants (*Acantholepsis fraunenfeldi* in November through January at W. Qilt and *A. bipartita* in October/November at Yarhiv). More detailed study is needed, but it is the author's impression that there may be a differential attractiveness of male and female flowers. The mirid plant bug mentioned above feeds on flowers and flower buds. There are a number of phytophagous (apparently polyphagous) thrip species found on Sumac flowers: *Retithrips syriacus* (Thripidae) and *Liothrips* sp. (Phlaeothripidae) are found in November at Yarhiv, *Odontothrips karyni* (Thripidae), *Haplothrips* sp. (Phlaeothripidae), and *Aeolothrips meridonalis* (Aeolothripidae) are found in December and January at W. Qilt.

Another rather important aspect of flowers involves the predatory spiders, especially crab spiders which sit on flowers or flower clusters waiting to feed on the various insects attracted to the flowers. The following crab spiders (Thomisidae) are found in association with *R. tripartita* and are more or less the same color as the flowers: *Synaema diana* at W. Qilt in November, *Thomisus* sp. at W. Qilt in December and January, *T. citrinellus* at Yarhiv in November, and *T. ornustus* at Yarhiv in October/November and probably December. At the peak season for flower visitors at Yarhiv (November), *T. ornustus* is very abundant yet difficult to spot visually (Fig. 15); various other more obviously colored spiders are around flower clusters and twigs during the flowering season at Yarhiv, but none as regularly as the two species of *Thomisus*.

Pollination of *R. tripartita* is primarily accomplished by a variety of small Diptera and Hymenoptera in both desert and Mediterranean environments, besides the flower and nectar feeders mentioned above which are all potential pollinators. Time did not allow detailed study of which Diptera, Hymenoptera, etc., species visit the flowers during the entire season and which ones regularly visit both male and female flowers.

Finally, classified as unnatural associates of *R. tripartita* are man and his domestic animals. Their effect is the most obvious in the populations of the Judean and Negev Deserts where Bedouin live off the sensitive desert ecosystem; taking but rarely giving back the vegetation. Besides the obvious occasional value of the wood from any tree in the desert, the main impact on the desert Sumac populations is intense grazing pressure by Bedouin animals, primarily goats and sheep (Fig. 3). This is most pronounced in a large population of Sumac such as that in W. Qilt. There the constant grazing, especially in the winter months, keeps most *R. tripartita* plants below one-half meter in height (Fig. 2). This is quite evident in precipitous, protected, or remote places not subjected to grazing, e.g., some single trees or small populations in the Negev or Sinai Mountain populations where these trees can reach heights of more than three meters (Fig. 1).

In the Mediterranean populations such as Yarhiv, grazing exists but is much less important because of a greater diversity of other food plants for domestic animals. However, in these areas man himself often destroys vast areas of *R. tripartita* by cutting it for firewood or for creating thorny protective barriers, or simply to clear the land for agricultural reasons.

Table 6. Fauna of *Rhus tripartita*—Flowers and Fruits

Species	Plant Part/Parasite	Mediterranean	Desert
Retithrips syriacus (Mayet)	flower	+	–
Liothrips sp.	flower	+	–
Aeolothrips meridionalis Priesner	flower	–	+
Odontothrips karnyi Priesner	flower	–	+
Haplothrips sp.	flower	–	+
Megastigmus pistaciae Walker	fruit	+	+
Mesopolobus sp.	larval parasite (*M.p.*)	+	–
Eupelmus urozonus Dalman	larval/hyper-(*M./M.p.*) parasite	+	–

– = absent
+ = present

Table 7. Fauna of *Rhus tripartita*—Associates

Species (Group)	Niche	Plant Part	Mediterranean	Desert
Stegodyphus sp. (Eresidae)	predator	branch	–	+
Steatoda maura (Simon) (Therididae)	predator	branch	–	+
dictynid sp. (Aranedia)	predator	twig	–	+
Eusparssus sp. (Sparassidae)	predator	branch	+	–
Mantis sp.	predator	branch	–	+
mantispid sp. (Neuroptera)	parasite	branch	–	+
Synaema diana (Audouin)	predator	flower	–	+
Thomisus sp.	predator	flower	–	+
Thomisus onustus Walckenaer	predator	flower	+	–
Acantholepis fraunenfeldi Mayr	nectar-feeder	flower	–	+
Acantholepis bipartita (Smith)	nectar-feeder	flower	+	–
micro-Diptera	pollinators	flower	+	+
micro-Hymenoptera	pollinators	flower	+	+

– = absent
+ = present

DISCUSSION

Phytogeography

The genus *Rhus* is apparently an ancient genus originating in the Cretaceous to early Tertiary when North America and Eurasia were still in contact (before the complete opening of the North Atlantic Ocean) but separated from Africa by the Tethys Sea until the Neogene (Quézel 1979 and *in litt.*). In the early Tertiary, equatorial climatic conditions existed in the southern Sahara. The Sahara went through climatic changes (from equatorial to tropical with dry seasons) during the Eocene-Oligocene transition; and in the Oligocene-Miocene, the tropical vegetation (adapted to dry seasons) existed there. By the lower Pliocene palynological evidence demonstrates that desert (xerophytic) conditions prevailed in this area (Maley 1980). In fact, Maley discovered lower Pliocene *Rhus* pollen in the Hoggar of the Sahara to be only a slighly higher percentage than at present. Along the Mediterranean coast of North Africa the Pliocene climate was evidently warmer and more humid than at present, but having dry summers (Quézel 1979). During the Pleistocene the climate in the Sahara fluctuated, with wet and dry phases as well as periods of colder temperatures. This brought about eliminations and exchanges of parts of different floras and created various regional mixtures of tropical Saharan and Mediterranean elements (Quézel 1979). Therefore, the genus *Rhus* probably spread from Eurasia to Africa in the early Tertiary, extended southwards throughout tropical Africa during the Pliocene and Pleistocene, and then speciated and radiated in different directions.

Possibly one of these radiations gave rise to *R. tripartita* and/or its relatives that spread from tropical African climates north again into the Sahara by Plio-Pleistocene times, and whose origin became obscured (relictual populations with apparently tropical phenology) by the vicissitudes of climate in Pleistocene North Africa and the adjacent Middle East. Therefore, *R. tripartita* may have a secondary tropical African origin, descended from a lineage of *Rhus* that invaded Central and Southern Africa originally from the northwest, adapted to a tropical environment, and then radiated north as equatorial climates were integrated with eremic phases, especially in the Pleistocene. *Rhus tripartita* was then (Pleistocene) adapted to a more moist environment but with periodic arid fluctuations, similar to the present-day Mediterranean climate. If this scenario for *R. tripartita* is correct, then actually it could be considered either a tropical African or Saharo-Arabian element of the Mesogean flora (*sensu* Quézel 1979), having evolved from tropical African ancestors but in the climatically shifting times of the Plio-Pleistocene of North Africa.

Zohary (1962, 1972, 1973) considers *R. tripartita* as a western Irano-Turanian element; however, even based on its present-day distribution, it seems unreasonable to consider this Sumac as Irano-Turanian. Previous studies of biogeographical affinities of some herbivorous beetles in the Middle East (Furth 1979), as well as bioclimatic, floristic, and population dynamic evidence for North African plants (Quézel 1979), do not indicate the presence of any significant Irano-Turanian element in the deserts of these regions. It is more likely that many plants of the high steppes and mountains of the Middle East and North Africa are tolerant (acclimatized) populations of present-day Mediterranean or Saharo-Arabian species.

There is some evidence that *R. tripartita* has affinities to a tropical African lineage. It has the trifoliate condition that all the east and south African *Rhus* have and which is not found in Palearctic species, e.g., *Rhus coriaria.* There is also evidence of chemical similarity of *R. tripartita* to tropical African *Rhus* species from studies of the secondary (defense) compounds, especially flavonoids (Furth and Young, in preparation). Furthermore, one of the major *R. tripartita* herbivores, the monophagous leaf beetle *Blepharida sacra,* has at least a partial biogeographic overlap with its food plant (it will not eat the common Palearctic Sumac *R. coriaria*) (Furth 1982), thus it has probably coevolved with *R. tripartita* and/or its tropical African relatives. This beetle's closest relatives are species that feed on mesophilic Sumac species in present-day tropical Africa (Furth, unpublished data). Therefore, this herbivore has apparently followed a similar biogeographic pattern and evolution as that suggested above for *R. tripartita,* that is, originating in tropical Africa and spreading [speciating] into the Sahara and Middle East, possibly in the pluvial Pleistocene. Unfortunately, not enough is known about this beetle's present-day or Pleistocene distributions. Apparently its populations are more sensitive to climatic change than its host and it can only survive where there are large stands of its food plant (Furth, unpublished data). Some other populations of this beetle may have been eliminated by climatic fluctuations in the Pleistocene. However, a biogeographical relationship (*sensu stricto*) is suggested for this herbivore and its Sumac food plant.

There are two basic environmental types of *R. tripartita* populations today. First, most common and widespread are the populations that live in desert environments under harsh xeric conditions across the North African Sahara (often in mountain locations) into the Middle Eastern deserts. These populations are usually quite small and are greatly isolated from the nearest population. Second are those populations living in a Mediterranean environment usually near the coast, e.g., SW Morocco, Algeria near Blidda, Canary Islands, Sicily, Linosa Island, Egypt, Israel, and Lebanon. Except for the island localities and Lebanon, these populations are not extremely remote from inland desert populations, but are isolated environmentally and ecologically from them. These Mediterranean populations (based on the field studies in Israel) involve densely populated stands of trees that dominate the maquis vegetation (Fig. 5).

I propose that these Mediterranean populations are, in fact, refugia—populations existing in a favorable environment similar to that in which the species flourished and was widespread in Pleistocene time. In this case, in the Pleistocene during the moist pluvials, *R. tripartita* was continuously widespread across North Africa into the Middle East. These wet pluvial phases had alternate dry phases but some probably had annual dry seasons similar to the present Mediterranean climate (Quézel 1979). In post-Pleistocene time, North Africa and the Middle East developed harsher xeric conditions that slowly eliminated vegetation needing a more moist environment. Most populations of *R. tripartita* in these desert environments began to shrink, leaving small isolated populations. These relict populations exist within the boundaries of their former widespread distribution in an environment that is different and more stressful than its original one. These populations are slowly but continuously shrinking. In Israel, the Mediterranean refugia and the desert relict populations are distinctly disjunct, with no trace of any link in the mountains of Samaria. It is possible that some connection between the Jordan Rift Valley and the Mediterranean Samaria populations of

R. tripartita existed even after the last pluvial and that the disjunction is relatively recent, but still prehistoric. There is no evidence or reason to support an idea of accidental or intentional transport of this plant by man from the desert to the Mediterranean area as suggested by Pollak (1983). As is often the case, there is little fossil evidence that sheds light on the past distribution of *R. tripartita.* However, Lorch's (1966) discovery of *R. tripartita* at Ubeidiya in the Jordan Valley near the southern end of the Sea of Galilee has been dated as early middle Pleistocene, during a moist pluvial period (Bar-Yosef and Tchernov 1972). The Ubeidiya site is considerably (ca. 50 km) north of present-day Jordan Valley populations of *R. tripartita* and the entire Jordan Rift Valley is currently a very xeric environment. The evidence is consistent with a more widespread distribution of *R. tripartita* in a pluvial/glacial Pleistocene period.

There are significant phenological differences between the desert and Mediterranean environment populations, yet the general winter-active seasonality is quite different from most of the other plant species in both areas. Pollak (1983) suggests that fall-winter activity in *R. tripartita* (beginning prior to the rainy season) implies a tropical origin (as with *Sisyphus* and *Ceratonia*) from a climate with late summer rain. Pollak (1983) rejects the possibility of its origination from a Mediterranean climate with early winter rains, but the evidence (above) suggests that refugial and relictual populations remain from widespread populations that flourished during pluvial periods of the Pleistocene in which wet and dry seasonal fluctuations differed from present Mediterranean ones. However, an apparently more ancient tropical African relationship and/or origin is evident from its morphology, chemistry, and herbivores; its phenology may also be remnant from this tropical connection. In any case, a true eremic (arid desert) origin seems improbable. Other tropical plants of the region (e.g., *Moringa, Maeura, Calotropis,* etc.) do not leaf, flower, and fruit at such an early season. A few other plants are active during the same fall-winter season and may be considered Mediterranean elements (e.g., *Urginea maritima*). Many species in the present-day Mediterranean and Sahara regions have rather obscured origins because climatic fluctuations of the Quarternary caused large scale floristic mixing; only some of these remain and often with varied phenology and ecology.

After studying the botanical, phenological, and ecological relationships of *R. tripartita,* it is apparent that its phytogeographical affinities are with an African continental source. It is evident that even though it is botanically related to species of present-day tropical Africa/Ethiopian Region (more closely than to the widespread Palearctic *R. coriaria*), its former broad distribution, at least in pluvial Pleistocene times, was in an environment similar to the modern Mediterranean (including much of the Sahara). Therefore, it is most appropriate to consider *R. tripartita* as a Pleistocene Mediterranean element whose relictual populations can marginally tolerate the quite xeric desert conditions.

Botany

Although *R. tripartita* requires calcareous soil, it manages to grow in a variety of edaphic situations, from rather deep clay-textured soil to fissures in hard limestone, dolomite rock, or granite, where some soil has accumulated. Wherever it grows it "creates" a high organic content soil from its senescent leaves and other debris. Also the chemistry of the senescent leaves (tannins, etc.) causes the soil to be a deep reddish-brown color and may have allelopathic qualities and/or -growth promoters for certain plants. Such properties have been found in the leaves and roots of other species of Sumac (Trotsenko 1959; Petranka and McPherson 1979; Matveev 1980), but this requires further soil biochemical investigations. Certainly an important reason for the ability of *R. tripartita* to produce leaves, flowers, and fruits before the rainy season and to subsist in the relictual populations of harsh arid environments is its capacity to retain significant amounts of water for a long time. Troitskaya and Mukhtarova (1961) demonstrated that *Cotinus coggygria* has a water retention capacity twice that of oak by decreasing its transpiration rate. Also, in order to survive in the xeric desert and even some Mediterranean environments, one would expect some reliance on the root system. Thus, it seems probable that water is retained better in the subterranean succulent rhizome system of *R. tripartita* than in the branches, leaves, and fruit. In a similar vein, it seems most conservative and efficient that propagation would be predominantly subterranean (vegetative reproduction), especially in the desert populations. Barkley (1937), referring to *Rhus* (*sensu lato*), says the root system is shallow and generally there is vegetative reproduction by rhizomes, populations often occurring in clumps or clones; however, individual specimens are common and normal in some species. Lyarskaya (1959) found that *R.*

typhina reproduces vegetatively via a surface root system that can radiate 15–16 meters from the parent tree and form adventitious buds that are dormant or form shoots; suckers sprout well and can be used for propagation. Studying *R. glabra* Lovell (1964) found that 60% of the fruit was nonviable due to wasp predation and that only 1–2% of unscarified fruits germinated; this increased three- or fourfold after burning. Lovell concluded that the Sumac colonies were primarily maintained by vegetative reproduction, especially in overgrazed and burned areas. This may be a significant factor in the maintenance of relictual desert populations. In the present study vegetative reproduction also seemed to be the most prevalent source of dispersal. Many small Sumac plants and shoots were dug up at both W. Qilt and Yarhiv and traced to a parent plant. Several attempts at germination of fruits were unsuccessful. Transplanted sprouts from both populations survived several months; however, proper facilities were unavailable for their continued maintenance and they eventually died. Small trees can be transplanted (J. Halperin, personal communication). In contrast, Pollak (1983) in laboratory experiments managed to get fruits to germinate and sprout in the first and second year (50% and 40%, respectively), but the 10 field sprouts he monitored died by summer. Pollak (1983) found no evidence of rhizoid reproduction and his twig rooting experiments failed, but there was good cut-stump regeneration of subterranean buds after cutting or fire. Distribution of male and female trees at Yarhiv and W. Qilt are probably not random, as suggested by Pollak (1983), but differential survival of rhizoid-spread trees may involve several factors. It seems likely that after the apparent high level of fruit infestation by *Megastigmus,* especially in the Mediterranean population, and the probable advantages of root dispersal in the desert populations, that vegetative reproduction is quite important and possibly the dominant mode for *R. tripartita.* Fruits may also be dispersed by frugivorous birds but no evidence or observations are yet recorded for this Sumac species. More extensive and carefully monitored studies are necessary in natural field situations for fruit germination versus rhizoid reproduction in Mediterranean and desert environments.

The difference in phenology between the Mediterranean and desert populations of *R. tripartita* is understandable considering the differences in precipitation, temperature regime, and soil types; basically two different environments. The seasonal differences are most notable for the beginning of the leaves, flowers, and fruits as well as for life cycles of insects associated with both populations, e.g., *B. sacra*. Even though such differences are real, they are rather difficult to quantify accurately because there is a significant range of variability at all stages. One interesting aspect of the ecological differences between the refuge and relict populations is deciduousness. As stated above, the Mediterranean population at Yarhiv is largely evergreen (most trees, especially large ones) although leaf production is very low in July and August; whereas the desert population at W. Qilt is completely deciduous from late June until early October. This also reflects the stressful xeric climatic conditions of the desert relict populations and the pressure for the conservation and retention of water for the following season. Fal'kova (1973) tested leaf cells in deciduous and evergreen Sumac species showing thermoresistance to be high in hot summer months for deciduous bushes and influenced by leaf dehydration in xerophytic plants, although in mesophytic plants he found no connection between thermoresistance and degree of leaf hydration. He concluded that solar radiation, duration of day, and plant age do not influence cellular thermoresistance in mature leaves. These facts are consistent with the pattern of deciduousness observed in the two populations of *R. tripartita*.

Several other botanical facts are noteworthy: There is a prevalence of red-brown color associated with several aspects of *R. tripartita*, i.e., new twigs, senescent leaves, heartwood, that is probably caused by the abundance of tannins and/or flavanoids. When the trunk is cut, there are no evident growth rings and, therefore, there is no accurate method for aging the trees (Furth, personal observation, and Pollak 1983). A significant number of Mediterranean and desert shrubs and trees are spinescent (Zohary 1962). Spinescence may be an adaptation to aridity or to grazing (Shmida and Whittaker 1981). The "pseudo-thorns" of *R. tripartita* are quite possibly adapted to grazing because only one other (*R. pentaphylla*) Sumac species in tropical Africa or the Palearctic has evolved spines or thorns, even though some also live in arid environments, e.g., *R. flexicaulis* Baker and *R. abyssinica* Hochst. Such an adaptation would presumably be rather recent, occurring after the advent of heavy grazing by domestic or wild browsers. Besides protection from large browsers, the spinescence provides protection and shelter to a variety of annual plants (e.g., Cruciferae, Boraginaceae, Compositae) and some animals (e.g., tenebrionid beetles, lizards, etc.). It is interesting that species of *Rhus* have vestigial anthers and stamens since this indicates a monoecious ancestor. Yet this phenomenon apparently predates the origin of *Rhus,* which probably took place in Laurasia in the late-Cretaceous; thus, if this condition is truly vestigial, it has been so for a very long time. This problem seems to merit further consideration.

Entomofauna

The arthropod herbivores of *R. tripartita* form an intricate system. This system contains herbivores that are restricted to this plant species (stenotopic or monophagous); herbivores that feed on a few other, often related plants (e.g., *Pistacia*) (oligophagous), and herbivores that are general feeders (polyphagous).

The most interesting of the herbivores are the monophagous ones, specifically linked to the *R. tripartita* system. These monophages have all coevolved with the secondary defense chemistry of this Sumac; tannins, flavinoids, and others. There are four or five species in this category that warrant special consideration. Although the mealybug, *Peliococcus deserticola,* was described as feeding on leaves of a borage early in the season (Furth, Ben-Dov, and Gerson 1984), it seems probable that the most obligate root-feeding stage of its life is restricted to *R. tripartita,* thus making it a monophage. The first host record of the leaf-rolling microlepidopteran, *Stenovalva diclusa,* was from larvae collected at the Yarhiv population and reared to adults. The only other records of this moth are from Jerusalem and the Zarqa River Colony in Jordan (Sattler 1982), both near known desert populations of *R. tripartita.* Thus, apparently this gelechiid can and does exist in both Mediterranean and desert Sumac populations but was very rare or remained at very low population levels in Judean and Negev desert populations during the four seasons of this survey, due to some unknown factor. The primary gelechiid found only in the desert populations is *Teliopsis sarcochroma* which is known only from *R. tripartita* and from a widspread distribution that parallels that of its host (Canary Islands across North Africa into the Middle East and Arabian Peninsula) (Sattler 1982). These two leaf-rolling gelechiids appeared to be filling the same niche and replacing each other (ecological vicars) in the Mediterranean and desert populations. An important monophage of *R. tripartita* is the processionary caterpillar, *Thau-*

metopoea jordana, that also has a distribution restricted to the desert environments in and adjacent to the Lower Jordan Rift Valley of Israel and Jordan. This moth will eat *Schinus terebinthifolius* (Furth and Halperin 1980), which indicates a chemical similarity between the genera. A close relative (*T. solitaria*) feeds on species of *Pistacia* (Davatchi 1958) and another (*T. cheela* Moore) on *Cotinus* (Furth and Halperin 1980). Finally, the Sacred Sumac Flea Beetle (*Blepharida sacra*), which can (in captivity) eat *Schinus* and some tropical African species of *Rhus* (Furth, unpublished data), is found abundantly in the Mediterranean population as well as in the large Judean Desert populations of *R. tripartita* and is only otherwise known from the desert environment near the Red Sea coast of Sudan (Furth 1982).

All five monophages are found in the relictual desert environment populations and only two in the refugia population of the Mediterranean environment. One might estimate that *B. sacra* and *S. diclusa* are older Sumac elements because only they have survived in the refuge population. However, their distribution at present is rather restricted, as it is for *T. jordana* and *P. deserticola,* implying possible relatively recent coevolution with their host species. In fact, *T. sarcochroma* may be the oldest of the *R. tripartita* guild because its distribution (biogeography) coincides so well with that of its host. Since the complete distribution and biology of *P. deserticola* is not yet certain, judgment about its restricted coevolution and biogeography with *R. tripartita* should be postponed. It seems most probable that *T. jordana* is the most recent Sumac monophage having evolved since the desertification of the climate in the Jordan Rift and possibly after the disjunction of relict and refuge populations. Apparently predating this climatic change and disjunction is *S. diclusa,* its present success in Middle East relictual populations being dubious. The beetle, *B. sacra,* also certainly evolved before the most recent Pleistocene aridification of the climate in N. Africa and the Middle East. However, because of its potential relation to the tropical African species of *Blepharida,* it may now have a remnant (relict) distribution of a species that thrived in a more tropical and widespread environment of the early Pleistocene or Pliocene. The most widespread monophage, *T. sarcochroma,* may also be old, however, if it is truly only found in relictual populations then it probably evolved in the late Pleistocene.

The oliphagous herbivores may also be somewhat restricted to groups of plants that have certain secondary chemicals in common. In this category one would expect to find *Pistacia* feeders whose distribution overlaps to some degree with *R. tripartita*. The seed wasp, *Megastigmus pistaciae,* is oliphagous on various species of *Pistacia* (as its name implies) in Mediterranean Europe and North Africa as well as in the Near and Middle East (Davatchi 1958). This wasp is an important herbivore of both Mediterranean and desert populations. The noctuid moth, *Eutelia adulatrix,* is well known on species of *Pistacia* and *Cotinus* in Mediterranean Europe, the Near East (Iran), and Central Asia (Davatchi 1958), with relatives on *Rhus* in North America, but is not really abundant on *R. tripartita* in either Mediterranean or desert populations. Because of its widespread distribution and rather restricted feeding, this moth (and relatives) probably has a long association with *Rhus*. The scale, *Melanaspis inopinatus,* although reported as polyphagous, especially on Rosaceae in Mediterranean Europe, appears to be primarily oliphagous in the Middle East on species of *Pistacia* (Davatchi 1958). Interestingly, in the present survey this scale was only found, and in large numbers, at the largest desert population (W. Qilt).

Most of the polyphagous herbivores on *R. tripartita* in either environmental population are usually unable to overcome the barriers of the other population's climatic zone and are, therefore, found only in one of those environments, e.g., *Imbecilla imbecilla,* common in the Mediterranean population, and *Zygina discolor,* common in the Judean Desert populations. Exceptions like *Lygus conspurcatus* are extremely common in the refugial Mediterranean populations and only rarely found in the relictual desert populations. In some cases these generalists fill complementary niches in each environment (ecological vicars), e.g., flower-feeding thrips, leaf-feeding leafhoppers. Others that may disperse easily like *O. tirhaca* are probably equally common in both environments. It seems probable that some polyphagous species such as *O. tirhaca* (see Davatchi 1958) may strongly prefer Anacardiaceous plants (*Pistacia, Rhus, Cotinus, Myrthus*) in the Middle East, but be capable of a more wide variety of food plants in other regions. There are many other polyphagous insects associated with *R. tripartita* on a rather casual basis that were not found with any regularity during this survey and, therefore, are omitted.

The arthropod herbivores of *R. tripartita* utilize most of the available resources and plant parts. Predictably the leaf-feeders are the largest group, yet more fruit-feeders might have been expected because this is a large potential food source. The number of monophagous and oliphagous herbivores associated with *R. tripartita* may seem small, but this may be due to a variety of factors, including plant secondary defense chemistry, climatic and phenological limitations of the two different types of populations, or geographical isolation with the relatively small population sizes especially in deserts. Referring to the study of the insect fauna of wild and cultivated species of *Pistacia* (Davatchi 1958), there are some categories and a few species of the oliphagous herbivores that may be expected to occur on *R. tripartita* (especially in the Mediterranean population); however, few of these would then be able to successfully colonize the desert populations of the Middle East. For example, *Pistacia* harbors many Homoptera such as aphids, psyllids, and scales even though most are polyphagous. Buprestidae, especially *Capnodis, Agrilus,* and *Anthaxia,* are prevalent on *Pistacia* and might be expected to attack *R. tripartita.* Chalcidoid wasps in the family Eurytomidae are reportedly rather common fruit feeder parasites on *Pistacia* and *Rhus* (see Appendix).

It is significant that surveys like this one usually discover smaller trophic systems within a major system, such as predator and parasite interactions (see Tables 4–7). Although not restricted to this plant, the general predators of the *R. tripartita* system (primarily spiders) are well supplied with a variety of herbivores as prey. Also the large diversity of insects, especially small Diptera and Hymenoptera, that are attracted to the flowers as nectar gatherers and/or pollinators provide prey for the predators. Crab spiders (Thomisidae) on the yellow-green Sumac flowers in both populations are well camouflaged in the visible spectrum. This has long been assumed to disguise these spiders from potential prey; however, they contrast markedly with the floral background when photographed in the UV light (Eisner et al. 1969). The reasoning is that crab spiders are not cryptic to their prey (insects) that see in the UV, but are cryptic to their potential predators (birds and lizards) that see in the visible spectrum. However, the situation is still not clear because there are some birds and lizards that can see in the UV and probably some insects that can see into the visible spectrum (G. Bernard, personal communication). Also, some crab spiders are able to change their visible color pattern and possibly alter this situation. In the *R. tripartita* system these crab spiders are very abundant during the peak season of herbivore and pollinator activity and, thus, it would seem unlikely that all these Sumac visiting insects are well aware of the presence of the crab spiders. One way to be more certain would be to take UV pattern photographs of both the crab spiders and *R. tripartita* flowers.

Conservation

The *R. tripartita* system is one of the best examples of grazing pressure by domestic animals in the deserts of the Middle East. In the largest desert population of this Sumac in Israel (W. Qilt and adjacent areas), the vast majority of the trees appear to be shrubs or bushes rarely over a meter high. This scrubby stature of the desert Sumac populations is caused by the continued grazing by the domestic goats, sheep, and sometimes camels that belong to the nomadic Bedouin people living in the Judean Desert (Figs. 2 and 3). This fact is quite evident, because in a few places at the edge of cliffs and other precipitous sites where the animals do not reach, the trees reach heights of over two meters, though these trees are still rather scrawny in appearance. Indeed, in W. Mukallik (a formal nature reserve) in the Judean Desert and in the Negev and Sinai populations, where there is no or minimal grazing, the trees are full and very broad and often three or more meters in height (Fig. 1). In some cases edaphic and/or special climatic conditions may stunt growth of a few trees; however, the contrast between grazed and ungrazed populations of *R. tripartita* is dramatic. The Samaria population (Mediterranean environment) is composed almost entirely of full broad trees averaging between two and three meters in height (Fig. 5). In the desert populations not only is the grazing pressure increased because of the scarcity of vegetation, but also because in the fall and early winter (October–December) *R. tripartita* is one of the only plants with a full complement of leaves, flowers, and fruits; therefore, a primary source of food for the grazers. However, the situation is quite different in the Mediterranean population because there are many more trees, shrubs, and other plants with leaves in the winter to disseminate any grazing pressure on *R. tripartita*. As mentioned previously, the thorny nature of this Sumac, like many other Mediterranean and desert plants, certainly saves it from complete eradication by domestic grazers, especially in the already dwindling relictual desert populations.

There are other direct human threats to the existence of this

Sumac, especially in the Samaria population, due to frequent cutting of many trees used for barricades around homes and domestic animals, in order to clear areas for agriculture, or just because its thorniness is a nuisance. Another effect, usually caused by man, is from fires, both accidental and intentional. Fortunately for its existence, Sumacs like *R. tripartita* regenerate fairly well from cut stumps and after fires (Lovell 1964, Pollak 1983).

In parts of the Mediterranean Samaria population, especially around Yarhiv, there may have been a slight reversal of the human effects on this Sumac due to political borders between Jordan (West Bank) and Israel from 1948 till 1967 and somewhat after. Being a border area there was probably less human development, i.e., settlements and agriculture, thereby decreasing the process of elimination of large numbers of *R. tripartita.* This localized trend is now rapidly being reversed again with the increase in population and resulting settlement, road building, and agriculture in that region.

Because of the various human-induced pressures on *Rhus tripartita* in both the Mediterranean and desert environments as well as the naturally stressful climatic conditions on the relictual desert populations, some form of conservation of this Sumac's ecosystem is necessary and timely. The only location where this Sumac species is within the boundaries of a formal nature preserve in Israel is in W. Mukallik (Nachal Ogg) where the significance of an ungrazed population is obvious from the larger tree size. None of the W. Qilt Sumac population is in the main formal natural preserve there, but only in the extension of the canyon west toward Jerusalem where domestic animal grazing on it is rampant. Also, none of the populations in the Negev Desert, that only consist of 1–10 trees each, is found in formal nature reserves. In some respects the most important population, the refugia in western Samaria, is not at all contained in a nature reserve (information on the current borders of nature reserves in Israel is taken from the five part, 1:100,000 maps of Israeli nature reserves, published by the Nature Reserves Authority of Israel in 1980). This Mediterranean population, even though still quite large, is the most vulnerable to expanding human development and agriculture; therefore, it is critical to preserve some of this partially ancient ecosystem before it is excessively partitioned and ruined. The population in and around W. Qilt is the largest in any of the desert regions of Israel and probably the largest such relictual population outside of the African continent. This population has been under great pressure from domestic grazing animals for a long time; it would be both revealing and rewarding to restore the natural aspect of this part of the Middle East desert ecosystem.

REFERENCES

Andrews, F. W. 1952. The flowering plants of Anglo-Egyptian Sudan. Vol. 2. Arbrouth, Scotland: T. Buncle & Co.

Ashbel, D. 1973. Climate, pp. 94–111. In Geography, Israel Pocket Library. Jerusalem, Keter Publishing House.

Balick, M. J., Furth, D. G., and Cooper-Driver, G. 1978. Biochemical and evolutionary aspects of arthropod predation on ferns. Oecologia (Berlin) 35:55–89.

Barkley, F. A. 1937. A monographic study of *Rhus* and its immediate allies in North and Central America, including the West Indies. Ann. Missouri Bot. Gard. 24(3):265–460.

Barkley, F. A. 1965. A criticism of the traditional concept of the genus *Rhus*. Prospects Iraq Biol. 3:52–58.

Barkley, F. and Barkley, E. 1938. A short history of *Rhus* to the time of Linnaeus. Amer. Midl. Natur. 19(2):265–333.

Bar-Yosef, O. and Tchernov, E. 1972. On the palaeo-ecological history of the site of 'Ubeidiya.' Jerusalem: The Israel Academy of Science and Humanities.

Brizicky, G. K. 1962. The genera of Anacardiaceae in the southeastern United States. J. Arnold Arbor. 43:359–375.

Brizicky, G. K. 1963. Taxonomic and nomenclatural notes on the genus *Rhus* (Anacardiaceae). J. Arnold Arbor. 44:60–80.

Danin, A. and Orshan, G. 1970. Distribution of indigenous trees in the Northern Central Negev highlands. La Ya'aran 20:91–98 (in Hebrew), 115–119 (English Abstract).

Danin, A., Orshan, G., and Zohary, M. 1975. The vegetation of the Northern Negev and the Judean Desert of Israel. Israel J. Bot. 24:118–172.

Davatchi, G. A. 1958. Étude biologique de la faune entomologique des *Pistacia* sauvages at cultivés. Rev. Pathol. vég. ent. Agric. France 37(1):3–166.

Diels, L. 1898. Die Epharmose der Vegetationsorgane bei *Rhus* L. & Gerontogeae Engl. Bot. Jahrb. 24:568–647.

Eisner, T., Silberglied, R., Aneshansley, D., Carrel, J., and Howland, H. 1969. Ultra-violet video-viewing: the televsion camera as an insect eye. Science 166:1172–1174.

Fal'kova, T. V. 1973. Seasonal changes in the thermo-resistance of higher plant cells under conditions of Mediterranean-type subtropics. Bot. Zhur. (Leningrad) 50(10):1424–1438 (in Russian).

Furth, D. G. 1979. Zoogeography and host plant ecology of the Alticinae of Israel, especially *Phyllotreta*: with descriptions of three new species (Coleoptera: Chrysomelidae). Israel J. Zool. 28(1):1–37.

Furth, D. G. 1980a. Zoogeography and host plants of *Longitarsus* in Israel, with descriptions of six new species (Coleoptera: Chrysomelidae). Israel J. Ent. 134:79–124.

Furth, D. G. 1980b. Wing polymorphism, host plant ecology, and biogeography of *Longitarsus* in Israel (Coleoptera: Chrysomelidae). Israel J. Ent. 13:125–148.

Furth, D. G. 1982. *Blepharida* biology, as demonstrated by the Sacred Sumac Flea Beetle (*B. sacra* (Weise)) (Coleoptera: Chrysomelidae: Alticinae). Spixiana Suppl. 7:43–52.

Furth, D. G., Ben-Dov, Y., and Gerson, U. 1984. A new species of *Pelicoccus* (Homoptera: Pseudococcidae) from the Judean Desert. Israel J. Ent. 17:105–108.

Furth, D. G., and Halperin, J. 1980. Observations of the phenology and biogeography of *Thaumetopoea jordana* (Staudinger) (Lepidoptera: Thaumetopoeidae). Israel J. Ent. 13:1–11.

Gillis, W. T. 1971. The systematics and ecology of Poison Ivy and Poison Oaks (*Toxicodendron*, Anacardiaceae). Rhodora 73:72–159, 161–237, 371–443, 465–540.

Israel Meteorological Service 1967. Climatological standard normals of rainfall, 1931–1960. Meterological Notes. Series A, no. 21.

Kingslover, J. M., Batra, S. W. T., and Utmar, J. A. 1984. A selected bibliography of insect-vascular plant associational studies. USDA, ARS, Bibliogr. Liter. Agric. No. 27.

LeHouerou, H. N. 1959. Recherches ecologiques et floristiques sur la végétation de la Tunisie meridionale, 2me Partie. Univer. Alger Inst. rech. Sahar. Mem. No. 6:1–219.

Lorch, J. 1966. A Pleistocene florulae from the Central Jordan Valley. Israel J. Bot. 15:31–34.

Lovell, J. F. 1964. An ecological study of *Rhus glabra* (an unpublished Ph.D. dissertation at Kansas State University). Dissertation Abstracts, 25(4):2190.

Lyarskaya, R. P. 1959. Certain specific features of the natural vegetative reproduction of *Rhus typhina* by means of root suckers. Nauch. Dokl. Vyss. Shk., Biol. Nauki 1:117–123 (in Russian).

MALEY, J. 1980. Les changements climatiques de la fin du Tertiaire en Afrique: leur conséquence sur l'apparition du Sahara et de sa végétation. In M. Williams and H. Favre, eds., The Sahara and the Nile. Rotterdam: A. A. Balkema.

MATVEEV, N. M. 1980. The rôle of leachates from the living woody plants in creation of allelopathic regime in forest stands of the steppe zone. Ekologiya 0(1):35–43.

MOUTERDE, P. 1970. Nouvelle flore du Liban et de la Syrie. Vol. 2 (texte). Beirut: Dar El-Machreq Editeurs.

NEGRE, R. 1962. Petite flore des region arides du Maroc Occidental. Vol. 2. Paris, Editions du Centre National de la Researche Scientifique.

PETRANKA, J. W. and McPherson, J. K. 1979. The role of *Rhus copallina* in the dynamics of the forest-prairie ecotone in north-central. Ecol. 60(5):956–965.

POLLAK, G. 1983. *R. tripartita* (Bernard da Ucria) Grande in Israel. Rotem, no. 6, pp. 17–31 (in Hebrew), pp. 74–75 (English summary).

POST, G. E. 1932. Flora of Syria, Palestine, and Sinai. Vol. 1. 2nd ed. (by J. R. Dinsmore). Beirut: American Press.

QUÉZEL, P. 1979. Analysis of the flora of Mediterranean and Saharan Africa. Ann. Missouri Bot. Gard. 65:479–534.

QUÉZEL, P. and SANTA, S. 1963. Nouvelle flore de l'Algerie. Vol. 2. Paris: Editions du Centre National de la Researche Scientifique.

RILEY, E. 1874. The jumping Sumach Beetle—*Blepharida rhois* (Forster). Missouri St. Ent., 6th Ann. Rep.:118–122.

SATTLER, K. 1982. A review of the western Palearctic Gelechiidae (Lepidoptera) associated with *Pistacia, Rhus,* and *Cotinus* (Anacardiaceae). Ent. Gaz. 33:13–32.

SHMIDA, A. and Whittaker, R. 1981. Pattern and biological microsite effects in two shrub communities in California. Ecol. 62(1):234–251.

TACKHOLM, V. 1974. Student's flora of Egypt. 2nd ed. Beirut: Cairo University.

TROITSKAYA, O. V. and Mukhtarova, N. Y. 1961. The water regime of *Cotinus coggygria.* Tr. Alma-Atinsk. Bot. Seda 6:119–125.

TROTSENKO, I. V. 1959. Sumac as an underbrush species. Lesn. Khoz. 5:30–32.

WILLIS, J. C. 1973. A dictionary of flowering plants and ferns. Revised by H. K. Airy Shaw. Cambridge: Cambridge University Press.

YOUNG, D. A. 1979. Heartwood flavonoids and the infrageneric relationships of *Rhus* (Anacardiaceae). Amer. J. Bot. 66(5):502–510.

ZOHARY, M. 1962. Plant life in Palestine. New York: Ronald Press.

ZOHARY, M. 1972. Flora Palaestina. Vol. 2: Plantanaceae to Umbelliferae. Jerusalem: The Israel Academy of Sciences and Humanities.

ZOHARY, M. 1973. Geobotanical foundations of the Middle East. Vol. 1. Stuttgart: G. Fischer Verlag.

ZOHARY, M., Heyn, C., and Heller, D. 1983. Fascicle 2, Conspectus Florae Orientalis. Jerusalem: The Israel Academy of Sciences and Humanities.

APPENDIX: ARTHROPODS OF RHUS, PISTACIA, AND COTINUS*

Taxa	*Hosts*	*Locality*	*Reference*
ACARINA			
ERIOPHYIDAE			
Aceria hexensis (Tucker)	*R.*sp.	S.Afr.	Tucker 1926
Aculops alachuae Keifer	*R.copallina*	FL	Keifer 1966
Aculops arizonarhus Keifer	*R.microphylla*	AZ	Keifer 1966
Aculops copallina (Keifer)	*R.copallina*	MD	Keifer 1959
Aculops morgani (Keifer)	*R.glabra*	Canada	Keifer 1966
Aculops rhoicecis (Keifer)	*R.trilobata*	CA	Keifer 1962a
Aculops toxicophagus (Ewing)	*R.diversiloba, R.radicans*	OR, CA	Ewing 1917
	R.radicans	OH, MD	Keifer et al. 1982
	R.vernix	OH	Keifer et al. 1982
Calacarus citrifolii Keifer (P)	*R.*sp.	S.Afr.	Jeppson et al. 1975
Epitrimerus typhinae Keifer	*R.typhina*	PA	Keifer 1962
Eriophyes convolvens (Nalepa) (P)	*R.glabra*	Eur., USA	Keifer 1938
Eriophyes giraldi Trotter	*R.*sp.	France	Trotter 1900
Eriophyes pistaciae Nalepa	*P.vera, P.terebinthus*	Italy, s.Eur.	Davatchi 1958
	P.vera	Iran	Davatchi 1958
Eriophyes rhoinus Cockerell	*R.*sp.	N.Amer.	Dodge and Rickett 1948
Eriophyes rhois Stebbins	*R.radicans*	IL	Stebbins 1910
Eriophyes stephanii Nalepa	*P.lentiscus, P.terebinthus*	USSR, c.Asia, s.Eur., n.Afr.	Davatchi 1958
	P.atlantica	Canary Is.	Davatchi 1958
	P.atlantica	Med., Asia Minor	Houard 1909
	P.mutica	USSR	Houard 1909
Heterotergum shevchenkoi Bagdasaryan	*P.mutica*	USSR (Armenia)	Bagdasaryan 1976
Phyllocoptes trilobi Keifer	*R.triloba*	CA	Keifer 1979
Tegonotus recki Bagdasaryan	*P.mutica*	USSR (Armenia)	Bagdasaryan 1976
TETRANYCHIDAE			
Brevipalpus californicus (Banks) (P)	*R.laevigata*	Cosmopolitan	Smith-Meyer 1979

*including some records for *Schinus,* excluding *R.tripartita* insects from the present study. See legend following Appendix for other keys to material.

Taxa	*Hosts*	*Locality*	*Reference*
Brevipalpus lilium Baker (P)	*R.*sp.	w.USA, FL, HI	Jeppson et al. 1975
Brevipalpus obovatus Donnadieu (P)	*R.leptodictya, R.undulata, R.longispina*	S.Afr.	Smith-Meyer 1979
Tenuipalpus lanceae Smith-Meyer	*R.lancea*	S.Afr.	Smith-Meyer 1979
Tenuipalpus mkuziensis Smith-Meyer (O)	*R.gueinzii*	S.Afr.	Smith-Meyer 1979
Tenuipalpus niekerkae Smith-Meyer (O)	*R.longispina*	S.Afr.	Smith-Meyer 1979
Tenuipalpus pyroides Smith-Meyer	*R.pyroides*	S.Afr.	Smith-Meyer 1979
Tenuipalpus rhusi Smith-Meyer	*R.laevigata, R.undulata*	S.Afr.	Smith-Meyer 1979
Tenuipalpus veckermanni Smith-Meyer (P)	*R.engleri*	S.Afr.	Smith-Meyer 1979
Tenuipalpus zuluensis Smith-Meyer (O)	*R.gueinzii*	S.Afr.	Smith-Meyer 1979
Tetranychus (Polynchus) schoeni McGregor (P)	*R.typhina*	e.se.USA	Jeppson et al. 1975
Tetranychus telarius Linnaeus (P)	*R.diversiloba*	OR	Ewing 1917
Tetranychus sp.	*P.vera*	Iran	Davatchi 1958
COLEOPTERA			
ANOBIIDAE			
Hedobia pubescens Olivier	*P.mutica, P.*sp.	USSR, c.Eur., Italy	Davatchi 1958
ANTHRIBIDAE			
Eusphyrus walshi LeConte (O)	*R.hirta*	PA	Champlain and Knull 1922
Tropideres curtirostris Mulsant & Rey (P)	*P.terebinthus, P.lentiscus*	USSR, s.Eur. n.Afr.	Davatchi 1958
ATTELABIDAE			
Attelabus analis Illiger (P)	*R.*sp.	N.Amer.	Felt 1906
Attelabus nigripes LeConte (M)	*R.*sp.	IN	Blatchley and Leng 1916
	R.aromatica, R.copallina, R.glabra	MO	J. Sullivan (pers. comm.)
Attelabus rhois Boheman (O)	*R.*sp.	e.USA	Blatchley and Leng 1916
Lasiorhynchites coeruleocephalus Schaller (P)	*P.lentiscus*	USSR, s.Eur. n.Afr.	Davatchi 1958
BOSTRICHIDAE (BORE IN DEAD WOOD)			
Scobicia chevrieri (Villa) (P)	*P.vera*	USSR, Iran, Syria, s.Eur.	Davatchi 1958
Xylodectus ornatus Lesne	*R.parviflora*	India	Mathur and Singh 1960

Taxa	*Hosts*	*Locality*	*Reference*
BUPRESTIDAE			
Acmaeodera despecta Baudi (P)	*P.vera*	Iraq	Knopf 1971
Acmaeodera judinae Stepanov	*P.vera*	USSR	Davatchi 1958
Agrilus derasofasciatus Lacordaire	*P.terebinthus*	USSR, Eur., n.Afr.	Davatchi 1958
Agrilus viridicerulans esfandiarii Obenberger	*P.vera, P.mutica*	Iran	Davatchi 1958
Agrilus sp.	*P.vera*	Aegean	Ulu et al. 1974
Anthaxia brevis Castelnau and Gory	*P.mutica*	USSR, Syria, Balkans	Davatchi 1958
	P.khinjuk	Iraq	Knopf 1971
Anthaxia ephippiata Redtenbacher	*P.mutica*	USSR, Iran	Davatchi 1958
Anthaxia judinae Stepanov	*P.vera*	USSR, Iran	Davatchi 1958
Anthaxia kreutzbergi Richter	*P.vera*	USSR	Davatchi 1958
Anthaxia millefolii (Fabricius) (P)	*P.lentiscus*	n.Afr.	Hellrigl 1978
Anthaxia umbellatarum Fab. (P)	*P.*sp., *R.pentaphylla*	Iran	Davatchi 1958
	P.lentiscus	Italy	Hellrigl 1978
	P.mutica	Near East	Hellrigl 1978
	P.terebinthus	France, Italy	Hellrigl 1978
	R.pentaphylla	n.Afr.	Hellrigl 1978
Capnodis anthracina Fisch.	*P.mutica*	Iran	Davatchi 1958
Capnodis carbonaria Klug	*P.*spp.	Iran	Davatchi 1958
	P.mutica	Iran	Hellrigl 1978
	P.vera	Syria	Hammad and Mohamed 1965
Capnodis cariosa (Pallas)	*P.vera*	Syria	Schneider 1958
	P.vera	Aegean	Ulu et al. 1974
	P.vera	Israel	Hellrigl 1978
	P.lentiscus	Italy, Yugo., USSR, Israel	Hellrigl 1978
	P.mutica, P.khinjuk, C.coggygria, Schinus molle	Iran	Hellrigl 1978
Capnodis cariosa hauseri Obenberger	*P.*spp.	Iran	Davatchi 1958
Capnodis cariosa iranica Bog. (P)	*P.vera*	Iran	Gentry 1965
Capnodis miliaris Klug	*P.vera*	Syria	Hammad and Mohamed 1965
Capnodis parumstriata Ball. (P)	*P.vera*	Iran	Davatchi 1958
Capnodis porosa Klug	*P.vera*	Syria	Hammad and Mohamed 1965
Capnodis tenebricosa Olivier	*P.vera*	Syria	Hammad and Mohamed 1965
Capnodis tenebrionis Linn.	*P.vera*	Syria	Hammad and Mohamed 1965
Chrysobothris affinis Linn. (P)	*P.vera*	Iraq	Knopf 1971
Chrysobothris jackovlevi Semenov	*P.*sp.	USSR	Davatchi 1958

Taxa	*Hosts*	*Locality*	*Reference*
CERAMBYCIDAE			
Astylidius parvus LeConte (P)	*R.radicans*	OH	Knull 1946
Astylopsis macula (Say) (P)	*R.radicans*	NC	Howden et al. 1951
Callimoxys sanguinicollis fuscipennis (LeConte) (P)	*R.diversiloba*	w.N.Amer.	Linsley 1963
Chlorophorus varius Mueller (P)	*R.vera*	Iraq.	Knopf 1971
Clytus monticola Gahan (O)	*P.integerrima*	Orient	Duffy 1968
Elaphidion mucronatum (Say) (P)	*R.*sp.	N.Amer.	Linsley 1963
Elaphidion villosum (Fab.) (P)	*R.glabra, R.typhina*	N.Amer.	Duffy 1960
Elaphidionoides villosus (Fab.) (P)	*R.*sp.	NY	Felt 1905
Erythrus westwoodi White	*R.semialata*	India	Mathur and Singh 1960
Eupogonius vestitus (Say) (P)	*R.radicans*	OH	Knull 1946
Hesperophanes sericeus (Fab.) (P)	*P.lentiscus*	USSR, Eur., n.Afr.	Davatchi 1958
Hyperplatys maculatus Haldeman (P)	*R.*sp.	e.N.Amer.	Felt 1906
Lampropterus ruficollis (LeConte)	*R.diversiloba*	w.N.Amer.	Linsley 1963
Leiopus alpha Say	*R.*sp.	N.Amer.	Felt 1906
Leiopus variegatus Haldeman (P)	*R.radicans*	OH	Knull 1946
Leptostylus albescens (Haldeman)	*R.radicans*	NC	Howden et al. 1951
Lepturges querci Fitch (P)	*R.radicans*	OH	Knull 1946
Lepturges signatus LeConte (P)	*R.radicans*	OH	Knull 1946
Macrotoma fisheri Waterhouse (P)	*P.chinensis*	Orient	Duffy 1968
Melanophila cuspidata (Klug) (P)	*P.lentiscus*	Italy	Hellrigl 1978
	P.vera	n.Afr., Italy, Greece, Syria	Gentry 1965
Molocrhus longicollis LeConte (P)	*R.*spp.	w.N.Amer.	Knull 1930
Nathrius brevipennis (Mulsant) (P)	*P.lentiscus*	s.Eur. (introduced n.Eur., n.Afr., w.N.Amer., S.Amer.)	Linsley 1963
Oberea ocellatus Haldeman	*R.*spp.	OH	Knull 1946
Oberea ruficollis Fab. (P)	*R.*sp.	N.Amer.	Felt 1906
Penichroa fasciata Stephens (P)	*P.lentiscus*	USSR, s.Eur., n.Afr.	Davatchi 1958
Perotis unicolor (Olivier) (P)	*P.lentiscus*	nw.Afr.	Hellrigl 1978
Poecilobrium chalybeum (LeConte) (P)	*R.*spp.	w.N.Amer.	Linsley 1963
Pogonocherus penicellatus LeConte	*R.*sp.	WV	Felt 1906

Taxa	*Hosts*	*Locality*	*Reference*
Prionus californicus Motschulsky (P)	*R.*spp.	w.N.Amer.	Linsley 1962
Saperda lateralis Fab. (P)	*R.radicans*	OH	Knull 1946
Saperda puncticollis Say	*R.radicans*	N.Amer.	Steyskal 1951
Stromatium barbatum (Fab.) (P)	*P.integerrima, R.insignis, R.punjabensis, R.wallichii*	India	Duffy 1968
Stromatium fulvum Villa (P)	*P.*sp.	USSR	Davatchi 1958
Xylotrechus quadripes Chevrolat (P)	*R.semiciliata*	Orient	Duffy 1968
CHRYSOMELIDAE			
Anomoea laticlavia (Foerster) (P)	*R.glabra*	MO	J. Sullivan (pers. comm.)
Antipa giguttata (Olivier)	*P.vera, P.lentiscus, P.terebinthus*	Greece	Tzanakakis 1962
Babia quadriguttata (Olivier) (P)	*R.glabra*	MO	J. Sullivan (pers. comm.)
Blepharida rhois (Foerster)	*C.coggygria, S.*sp.	USA	Frost 1973
	R.typhina, R.glabra, R.copallina, R.aromatica	N.Amer.	Riley 1874
	C.obovatus	MO	J. Sullivan (pers. comm.)
Blepharida marginalis (Weise)	*R.natalensis, R.vulgaris*	Kenya	Furth 1985
Blepharida sacra (Weise)	*R.tripartita*	Israel	Furth 1982
Blepharida sp.	*R.*sp.	S.Afr.	Scherer 1973
Blepharida sp.	*R.natalensis*	Kenya	Furth (unpubl.)
Chlamisus foveolatus Knoch	*R.radicans*	e.N.Amer.	Wilcox 1979
Clytra atraphaxidis (Pallas)	*P.vera*	Greece	Tzanakakis 1962
Clytra novempunctata (Olivier)	*P.vera, P.terebinthus*	Greece	Tzanakakis 1962
Coscinoptera dominicana Fab.	*R.*sp., *R.radicans*	e.N.Amer.	Wilcox 1979
Coscinoptera d. franciscana LeConte (P)	*R.copallina, R.glabra*	MO	J. Sullivan (pers. comm.)
Cryptocephalus fulvus Goeze	*P.vera, P.lentiscus*	Greece	Tzanakakis 1962
Cryptocephalus ilicis Olivier	*P.vera, P.lentiscus, P.terebinthus*	Greece	Tzanakakis 1962
Cryptocephalus imperialis Laicharting	*P.terebinthus*	Greece	Tzanakakis 1962
Cryptocephalus notatus Fab.	*R.*sp.	e.N.Amer.	Wilcox 1979
Cryptocephalus quadruplex Newman (P)	*R.glabra*	e.N.Amer.	Wilcox 1979
	R.aromatica, R.glabra	MO	J. Sullivan (pers. comm.)
Diachus chlorizans Suffrian	*R.*sp.	e.N.Amer.	Wilcox 1979
Gynanropthalma unipunctata (Olivier) (P)	*P.vera*	Iraq	Knopf 1971
Lachnaea orientalis Weise	*P.vera, P.lentiscus, P.terebinthus*	Greece	Tzanakakis 1962
Lachnaea variolosa Linn.	*P.lentiscus*	n.Afr.	Davatchi 1958
Lexiphanes affinis Haldeman	*R.copallina*	e.N.Amer.	Wilcox 1979
Macrolenes dentipes (Olivier)	*P.vera, P.lentiscus, P.terebinthus*	Greece	Tzanakakis 1962
Myopristis bimaculata Rossi	*P.lentiscus, R.coriaria*	n.Afr.	Davatchi 1958
Nodonota puncticollis Say (P)	*R.*sp.	WV	Felt 1906

Taxa	*Hosts*	*Locality*	*Reference*
Orthaltica copalina (Fab.)	*R.copallina, R.radicans*	e.N.Amer.	Wilcox 1979
	R.glabra	MO	J. Sullivan (pers. comm.)
	R.glabra	SD	Scherer 1974
Orthaltica melina Horn	*R.copallina*	MO	J. Sullivan (pers. comm.)
	R.glabra	SD	Scherer 1974
	R.radicans	PA	Scherer 1974
Orthaltica reticolis (LeConte)	*R.*sp.	CA	Scherer 1974
Orthaltica parkeri (B. E. White)	*R.*sp.	AZ	Scherer 1974
Pachybrachis anoguttatus Suffrian	*P.lentiscus*	Med., n.Afr.	Davatchi 1958
Pachybrachis limbatus (Menetries)	*P.vera, P.lentiscus, P.terebinthus*	Greece	Tzanakakis 1962
Pachybrachis simius Marsham	*P.lentiscus*	Algeria	Davatchi 1958
Pachybrachis tridens Melsheimer (P)	*R.radicans, R.*sp.	e.N.Amer.	Wilcox 1979
Podontia affinis Groendal (O)	*R.*sp.	?Taiwan	Takizawa 1978
Podontia lutea (Olivier) (M)	*R.succedanea*	Taiwan	Takizawa 1978
Podontia quatuordecimpunctata (Linn.) (O)	*R.*sp.	?Taiwan	Takizawa 1978
Rhaphidopalpa foveicollis (Lucas) (O)	*P.vera*	Iran	Davatchi 1958
Serraticollis parkeri White	*R.*sp.	AZ	White 1942
Serraticollis rhois White	*R.*sp.	CA	White 1942
CURCULIONIDAE			
Cryptorhynchus fuscatus LeConte	*R.radicans*	NC	Howden et al. 1951
Cryptorhynchus obliquus Say	*R.*sp.	WV	Felt 1906
Cylindrocopturus binotatus (LeConte) (P)	*R.*sp.	NY	Felt 1906
Lignyodes enucleator Panzer	*P.*spp.	USSR	Davatchi 1958
Madarellus undulatus Say (P)	*R.toxicodendron*	NJ	Blatchley and Leng 1916
Polydrosus davatchii Hoffmann	*P.vera*	Iran	Davatchi 1958
Polydrosus elegantus Boheman	*P.lentiscus*	Greece, Yugo., Sicily	Davatchi 1958
Pseudocoeliodes rubricus (Gyllenhal)	*P.vera*	Greece	Tzanakakis 1962
Strophosomus lentisci Hoffman	*P.lentiscus*	Morocco	Davatchi 1958
DERMESTIDAE			
Apsectus hispidus (Melsheimer)	*R.radicans*	e.USA	Hicks 1955
ELATERIDAE			
Ctenicera hamata Say	*R.toxicodendron*	MD	Steyskal 1951

Taxa	*Hosts*	*Locality*	*Reference*
MELANDRYIDAE			
Synchroa punctata Newman	*R.toxicodendron*	PA	Knull 1930
RHYNCHITIDAE			
Eugnamptus punctipes LeConte	*R.copallina*	MD	Blatchley and Leng 1916
SALPINGIDAE			
Lissodema lituratum Costa	*P.lentiscus*	Corsica	Davatchi 1958
SCOLYTIDAE			
(Because of the complicated synonymy in this family, if the genus and/or species name given in the original reference differs from the current name, according to Wood (1982), the original name will be included in parentheses following the specific name. Names of synonymous species are bracketed.)			
Cactopinus rhois Blackman (M)	*R.diversiloba, R.ovata, R.trilobata*	CA	Wood 1982
Chaetophloeus hystrix (LeConte) (M)	*R.integrifolia, R.laurina, R.ovata*	s.CA	Wood 1982
Chaetophloeus lasius Wood	*R.*sp.	Mex.	Wood 1982
Chaetophloeus penicillatus (Bruck) (*Renocis*)	*R.*spp.	CO, UT, AZ, CA, Mex.	Blackman 1940
	R.integrifolia	CA	Bruck 1933
	*R.integrifolia, R.microphylla, R.ovata, R.trilobata, R.*sp.	AZ, CA, CO, NV, NM, UT, Mex.	Wood 1982
Cryptocarenus seriatus Eggers (P)	*R.(Metopium) toxiferum*	S./C./N. Amer.	Wood 1982
Estenoborus perrisi Chapuis (O)	*P.vera, P.mutica*	Iran	Davatchi 1958
	P.terebinthus, P.lentiscus	USSR, Turkey, Greece, Bulg., Med.Eur./Afr.	Davatchi 1958
	P.atlantica	n.Afr.	Davatchi 1958
Hylesinus vestitus (Mulsant and Rey) (*Chaetoptelius*)	*P.vera, P.*spp.	USSR, Yugo., s.Eur., n.Afr.	Davatchi 1958
	P.vera	Aegean	Ulu et al. 1974
	P.vera	Syria	Schneider 1958
	P.vera	Iraq	Rizk and Ardini 1981
	P.mutica	Iran	Davatchi 1958
Hylocurus hirtellus (LeConte) (P)	*R.*sp.	w.USA	Wood 1982
Hypothenemus aspericollis Wollaston (P)	*P.*sp.	France	Davatchi 1958
Hypothenemus distinctus Wood	*R.aromatica*	MO	Wood 1982
Hypothenemus hispidulus Leconte	*R.*sp.	NJ	Blatchley and Leng 1916
Hypothenemus hirsutus (Wood) (P)	*R.(Metopium) toxiferum*	FL	Wood 1982
Hypothenemus indigens Wood (P)	*Toxicodendron* sp.	Mex.	Wood 1982
Hypothenemus interstitialis (Hopkins) (P)	*R.*spp.	e.N.Amer.	Wood 1982

Taxa	*Hosts*	*Locality*	*Reference*
Hypothenemus lezhavai Pjatnitsky (P)	*P.mutica*	USSR	Davatchi 1958
Hypothenemus seriatus (Eichoff) (P) [*Stephanoderes georgiae* Hopkins]	*R.glabra*	NC	J. M. Schmid (pers. comm.)
Lymantor decipiens (LeConte) (P)	*R.typhina*	e.N.Amer.	Wood 1982
Micracisella nigrella Wood	?*R.*sp.	Guat.	Wood 1982
Pityophthorus concentralis Eichoff	*R.metopium*	FL	Blatchley and Leng 1916
Pityophthorus corruptus Wood	*Toxicodendron* sp.	Mex.	Wood 1982
Pityophthorus crinalis Blackman	*R.(Tox.)vernix*	e.USA	Wood 1982
Pityophthorus detentus Wood	*Toxicodendron* sp.	Mex.	Wood 1982
Pityophthorus hylocuroides Wood	*R.?aromatica*	Mex.	Wood 1982
Pityophthorus lautus Eichoff (P) [*Pityophthorus rhois* Swaine]	*R.typhina*	NY	Loan and Matthews 1973
	R.glabra	NC	J. M. Schmid (pers. comm.)
	R.radicans	NC	Howden et al. 1951
Pityophthorus scriptor Blackman	*R.typhina*	se.USA	Wood 1982
Pityophthorus virilis Blackman	*R.trilobata*	w.USA	Wood 1982
Stenocleptus sulcatus (Bruck) (O)	*R.integrifolia, R.laurina*	s.CA	Wood 1982
Thysanoes inornatus Wood (O)	*R.*sp.	Mex.	Wood 1982
Trischidias atoma (Hopkins) (P) [*Hypothenemus toxicodendri* Hopkins]	*R.radicans*	WV	Blatchley and Leng 1916
Xyleborus affinis Eichhorn (P)	*R.trilobata*	SD	J. M. Schmid (pers. comm.)
[*Pityophthorus consimilis* LeConte]	*R.glabra, T.toxicodendron*	e.N.Amer.	Blatchley and Leng 1916
Xyleborus saxeseni (Ratzeburg) (P)	*P.mutica*	USSR	Davatchi 1958
[*Xyleborus pecanus* Hopkins]	*R.radicans*	NC	Howden et al. 1951
Xyleborus tachygraphus Zimmerman (*Ambrosiodmus*) (P)	*R.*sp.	PA, NY, mid./ s.USA	Blatchley and Leng 1916
Xylosandrus germanus (Blandford) (P)	*R.radicans, R.typhina*	USA	Weber and McPherson 1983
	R.ambigua, R.chinensis, R.succedanea, R.verniciflua	Japan	Weber and McPherson 1983
	R.sylvestris	Korea	Weber and McPherson 1983

Taxa	*Hosts*	*Locality*	*Reference*
DIPTERA			
CECIDOMYIIDAE			
Asphondylia integrifoliae Felt	*R.integrifolia, R.trilobata*	CA, UT	Pritchard 1953
Contarinia cotini Kieffer	*C.coggygria*	Eur.	Barnes 1948
Dasyneura cotini Janezic	*C.coggygria*	Yugo.	Janezic 1978
Dasyneura rhois (Coquillet)	*R.toxicodendron*	NY	Barnes 1948
Parallelodiplosis cinctipes Felt	*R.vernix*	MA	Barnes 1948
CHAMAEMYIIDAE			
Leucopis palumbi Rondani (ex aphid galls)	*P.lentiscus*	France	Davatchi 1958
EMPIDIDAE (POLLINATORS)			
sp.	*R.aromatica*	IL	Robertson 1896
LONCHAEIDAE (POLLINATORS)			
sp.	*R.aromatica*	IL	Robertson 1896
MUSCIDAE (POLLINATORS)			
sp.	*R.aromatica*	IL	Robertson 1896
SARCOPHAGIDAE (POLLINATORS)			
sp.	*R.aromatica*	IL	Robertson 1896
SCIOMYZIDAE (POLLINATORS)			
sp.	*R.aromatica*	IL	Robertson 1896
SYRPHIDAE (POLLINATORS)			
3 spp.	*R.aromatica*	IL	Robertson 1896
TACHINIDAE (POLLINATORS)			
sp.	*R.aromatica*	IL	Robertson 1896
TEPHRITIDAE			
Rhagoletis fausta (Osten Sacken)	*Rhus typhina*	WI	Prokopy 1976
HEMIPTERA			
HETEROPTERA			
ALYDIDAE			
Megalotomus quinquespinosus (Say) (P)	*R.glabra*	MI	Schaefer 1980
ANTHOCORIDAE (PREDATORS OF GALL APHIDS = FORDINAE)			
Anthocoris gallarum-ulmi (DeGeer)	*P.lentiscus, P.terrebinthus*	Med.	Davatchi 1958
Anthocoris pericus Wagner (on *Asiphonella cynodonti*)	*P.khinjuk*	Iran	Davatchi 1958
Anthocoris minki pistaciae Wagner (on *A. cynodonti, Forda*)	*P.vera*	Iran	Davatchi 1958

Taxa	*Hosts*	*Locality*	*Reference*
Orius horvathi Reuter (on *A. cynodonti*)	*P.khinjuk*	Iran	Davatchi 1958
COREIDAE			
Gonocerus acuteangulatus Goeze	*P.*sp.	Sicily	Davatchi 1958
Gonocerus venator Fallen	*P.vera*	Sicily	Davatchi 1958
LYGAEIDAE			
Graptostethus servus (Fab.)	*P.vera*	Sicily	Romeo 1929
Heraeus plebejus Stal	*R.*sp.	e.USA	Blatchley 1926
Lygaeus kalmii Stal	*R.glabra*	IL	Robertson 1929
Lygaeus turcicus Say	*R.glabra*	IL	Robertson 1929
	R.hirta	IN	Blatchley 1926
Plinthisus cautus Popov	*P.vera*	Iran, USSR	Popov 1965
Spilostethus pandurus (Scopoli)	*P.vera*	Sicily	Davatchi 1958
MIRIDAE			
Creontiades elongatus Lethierry	*R.vulgaris*	Uganda	LePelley 1959
Lygocoris communis (Knight)	*R.*sp.	Quebec	Boivin and Stewart 1982
Taylorilygus simonyi Reuter	*R.natalensis*	Uganda	LePelley 1959
Taylorilygus vosseleri Poppius	*R.natalensis*	Uganda	LePelley 1959
Neurocolpus nubilus (Say)	*R.typhina*	w.Can.	Kelton 1980
Poecilocapsus lineatus (Fab.) (P)	*R.*sp.	N.Amer.	Johnson and Lyon 1976
Prepops fraternus (Knight) (O)	*R.aromatica*	IL	Knight 1941
PENTATOMIDAE			
Apodiphus amygdali (Germar) (P)	*P.mutica*	Near East	Kirkaldy 1909
Carpocoris pudicus Poda	*P.vera*	Sicily	Davatchi 1958
Codophila varia Fab.	*P.vera*	USSR	Davatchi 1958
Graphosoma lineatum Linn.?	*P.vera*	Sicily	Davatchi 1958
SCUTELLERIDAE			
Stethaulax marmoratus (Say)	*R.glabra, R.copallina, R.aromatica*	IL	Walt and McPherson 1973
TINGIDAE			
Corythucha padi Drake (O)	*R.toxicodendron*	CO, New Bruns.	Drake and Ruhoff 1965
Elasmotropis testacea (Herrich-Schaeffer) (O)	*P.*sp.	Cyprus, ?Mid/ Near East	Drake and Ruhoff 1965
Teleonemia nigrina Champion (P)	*R.laurina*	USA	Drake and Ruhoff 1965
Tingis hellenica (Puton) (O)	*P.*sp.	Cyprus	Drake and Ruhoff 1965
HOMOPTERA			
ALYEURODIDAE			
Acaudaleyrodes citri (Priesner & Hosny) (P)	*R.*sp.	trop. & n.Afr., Mid-East, India	Mound and Halsey 1978

Taxa	*Hosts*	*Locality*	*Reference*
Aleurodicus coccolobae Quaintance & Baker	*S.terebinthefolius*	Neotropics	Mound and Halsey 1978
Aleurodicus dispersus Russell (P)	*S.terebinthefolius*	Neotropics, FL, Canary Is.	Mound and Halsey 1978
Aleuroparadoxus punctatus Quaintance & Baker	*S.*sp.	Chile	Mound and Halsey 1978
Aleurothrixus porteri Quaintance & Baker (P)	*S.molle, S.despondens*	Argen., Braz., Chile	Mound and Halsey 1978
Asterobemisia carpini (Koch) (P)	*C.coggygria*	Eur.	Mound and Halsey 1978
Bemisia silvatica Danzig (P)	*C.coggygria*	USSR	Mound and Halsey 1978
Pealius rubi Takahashi (O)	*R.*sp.	Japan, Taiwan	Mound and Halsey 1978
Trialeurodes intermedia Russell (O)	*R.ovata, R.*sp.	CA, TX	Mound and Halsey 1978
Trialeurodes packardi (Morrill) (O)	*R.aromatica*	USA	Mound and Halsey 1978
Trialeurodes varporariorum (Westwood) (P)	*S.molle*	?CT	Mound and Halsey 1978
	R.diversiloba, R.virens	N.Amer.	Mound and Halsey 1978

APHIDIDAE

(Because of the complicated synonymy in this family, if the genus and/or species name given in the original reference differs from the current name, according to Eastop and Hille Ris Lambers (1976), the original name will be included in parentheses following the specific name. Names of synonymous species are bracketed.)

Aphis chinensis Doubleday	*R.coriaria*	Syria	Houard 1909
Aphis gossypii Glover (P)	*P.*spp.	Iran, n.Afr.	Davatchi 1958
Aphis rhoicola Hille Ris Lambers	*R.abyssinica*	Ethiopia	Richards 1976
Aphis therebinthinae Virey	*P.terebinthus*	?	Patch 1938
Aploneura lentisci Passerini	*P.lentiscus*	Iran, Med., Eur.–Asia	Davatchi 1958
	P.lentiscus	Israel	Wertheim 1954
Aploneura sp.	*P.atlantica*	Israel	Wertheim 1954
Asiphonella cynodonti (Das) (O)	*P.khinjuk*	Iran, s.India, Egypt	Davatchi 1958
[*Asiphonella dactylonii* Theobald]	*P.khinjuk, P.sinensis*	Iran, Egypt	Davatchi 1958
Aulacorthum sp.	*P.vera*	Iran	Davatchi 1958
Baizongia pistaciae (Linn.) (O)	*P.khinjuk, P.terebinthus*	circum-Med.–n.India	Davatchi 1958
	P.palaestina	Israel	Wertheim 1954
[*Pemphigella aedificator* Buckton]	*P.chinensis, P.khinjuk, P.integerrima, P.terebinthus*	?India, Formosa	Patch 1938
[*Pemphigella cornicularia* Passerini]	*P.chinensis, P.khinjuk*	?USSR	Patch 1938

Taxa	*Hosts*	*Locality*	*Reference*
	P.terebinthus, P.vera	s.Eur.	Patch 1938
	P.lentiscus, P.atlantica, P.narbonensis (fossil)	Cyprus	Houard 1909
[*Pemphigus corniculoides* Lichtenstein]	*P.terebinthus*	?France	Patch 1938
Baizongia sp. A	*P.palaestina*	Israel	Koach and Wool 1977
Baizongia sp. B.	*P.palaestina*	Israel	Koach and Wool 1977
Baizongia sp. C.	*P.palaestina*	Israel	Koach and Wool 1977
Carolinaia rhois Tissot	*R.radicans*	OH	Tissot 1928
Chaetogeoica foliodentata (Tao)	*P.sinensis*	China, Taiwan	Davatchi 1958
Chuansicallis chengtuensis Tao	*R.javanica*	China	Richards 1976
Forda formicaria von Heyden (O)	*P.terebinthus*	Med.	Davatchi 1958
	P.palaestina	Israel	Koach and Wool 1977
[*Pemphigus semilunarius* Passerini]	*P.lentiscus, P.atlantica*	Med.	Houard 1909
	P.vera	e.Eur., Sicily	Houard 1909
	P.palaestina	Israel	Wertheim 1954
[*Pemphigus semilunoides* Lichtenstein]	*P.terebinthus*	?France	Patch 1938
Forda hirsuta Mordvilko	*P.vera, P.mutica*	Iran	Davatchi 1958
	P.vera	USSR	Patch 1938
Forda kaussarii Davatchi & Remaudiere	*P.khinjuk*	Iran	Davatchi 1958
Forda marginata Koch (O)	*P.terebinthus*	Iran, Eur.	Davatchi 1958
[*Forda follicularius* (Passerini)]	*P.vera, P.mutica, P.terebinthus*	s.Eur.	Patch 1938
	P.atlantica	Med.	Houard 1909
	P.palaestina	Israel	Wertheim 1954
[*Forda folliculariodes* Mordvilko]	*P.terebinthus*	s.Eur.	Patch 1938
[*Pemphigus folliculoides* Lichtenstein]	*P.terebinthus*	?France	Patch 1938
[*Pemphigus retroflexus* Courchet]	*P.atlantica*	Med.	Houard 1909
Forda rotunda Theobald [*Fordona italica* Mordvilko]	*P.*sp.	Italy	Richards 1976
[*Pentaphis trivialis* (Passerini)]	*P.terebinthus, P.mutica*	Balkans–USSR	Patch 1938
	P.lentiscus	?Eur.	Patch 1938
Forda riccobonii (Stefani)	*P.atlantica*	Morocco	Davatchi 1958
	P.atlantica	Sicily, Canary Is.	Houard 1909
	P.mutica	Asia Minor	Houard 1909
	P.atlantica	Israel	Koach and Wool 1977
Forda sichangensis Remaudiere & Tao	*P.sinensis*	?Iran	Davatchi 1958

Taxa	*Hosts*	*Locality*	*Reference*
[*Pemphigella marginata* Tao]	*P.sinensis*	w.China	Richards 1976
Fordini sp. A	*P.atlantica*	Israel	Koach and Wool 1977
Fordini sp. B	*P.atlantica*	Israel	Koach and Wool 1977
Geoica lucifuga Zehntner (P)	*P.chinensis*	?USSR	Patch 1938
Geoica mimeuri (Gaumont)	*P.atlantica, P.terebinthus*	Morocco	Davatchi 1958
Geoica muticae (Mordvilko)	*P.mutica*	Iran	Davatchi 1958
[*Pemphigetum muticae* Mordvilko]	*P.mutica*	?s.Eur.	Patch 1938
[*Geoica utricularia rungsi* Davatchi & Remaudiere]	*P.atlantica*	Morocco	Davatchi 1958
Geoica setulosa (Passerini)	*P.khinjuk*	Iran	Davatchi 1958
Geoica utricularia (Passerini)	*P.terebinthus*	Iran	Davatchi 1958
	P.khinjuk	Egypt (Sinai)	Koach and Wool 1977
	P.atlantica, P.palaestina	Israel	Koach and Wool 1977
[*Pemphigus utriculoides* Lichtenstein]	*P.terebinthus*	?France	Patch 1938
Glabromyzus rhois (Monell) (*Rhopalosiphum*)	*R.canadensis, R.glabra, R.diversiloba, R.trilobata*	N.Amer.	Patch 1938
Glabromyzus rhusfoliae (Richards) (*Aulacorthum*)	*R.radicans*	Ontario	Richards 1972
Glabromyzus schlingeri Hille Ris Lambers	*R.diversiloba*	CA	Richards 1976
Juncomyzus rhois (Takahashi) (*Myzus*)	*R.vernicifera*	Taiwan	Patch 1938
Kaburagia rhusicola Takagi (*Macrorhinarium ensigallis* Tsai & Tang) (*Macrorhinarium ovogallis* Tsai & Tang)	*R.punjabensis sinica*	China	Tsai and Tang 1946
Melaphis rhois (Fitch)	*R.glabra, R.trifolia, R.typhina*	N.Amer.	Patch 1938
Nurudea ibofushi Matsumura	*R.semialata*	Japan	Patch 1938
[*Nurudea sinica* Tsai & Tang]	*R.javanica*	China	Tsai and Tang 1946
Nurudea meitanensis (Tsai & Tang) (*Floraphis*)	*R.punjabensis sinica*	China	Tsai and Tang 1946
Nurudea shiraii (Matsumura)	*R.semialata*	Japan	Patch 1938
	R.javanica	Formosa	Patch 1938
Nurudea yanoniella (Matsumura)	*R.semialatu*	Japan	Patch 1938
[*Fushia rosea* Matsumura]	*R.semialata*	Japan	Patch 1938
Paracletus cimiciformis v. Heyden	*P.terebinthus, P.khinjuk*	Iran	Davatchi 1958
[*Paracletus palidus* (Derbes)]	*P.palaestina*	Israel	Wertheim 1954
[*Paracletus portshinskyi* Mordvilko]	*P.sp.?*	?Sicily	Patch 1938
[*Pemphigus derbesi* Lichtenstein]	*P.terebinthus, P.vera*	?Eur.	Patch 1938
	P.atlantica	Med.	Houard 1909

Taxa	*Hosts*	*Locality*	*Reference*
[*Pemphigus pallidoides* Lichtenstein]	*P.terebinthus*	?France	Patch 1938
Paracletus donisthorpei Theobald	*P.*sp.?	?Sicily	Patch 1938
Pemphigella foliodentata Tao	*P.sinensis*	w.China	Richards 1976
Pemphigus coccus Buckton	*P.vera*	?Afghan.	Patch 1938
Pemphigus minor Derbes	*P.terebinthus*	France	Patch 1938
Phyllaphis fagi (Linn.) (O)?	*P.vera*	Iran	Davatchi 1958
Prociphilus bumeliae (Schrank)	*R.semialata*	Japan	Patch 1938
Rectinasus buxtoni Theobald (P)	*P.khinjuk*	Iran	Davatchi 1958
Schlechtendalia chinensis (Bell)	*R.chinensis*	China	Tang 1976
	R.semialata	China	Patch 1938
	R.semialata	Japan	Monzen 1929
	R.japonica	Japan	Baker 1917
[*Schlechtendalia intermedia* Matsumura]	*R.semialata*	Japan	Patch 1938
[*Schlechtendalia mimifushi* Matsumura]	*R.semialata*	Japan	Patch 1938
[*Schlechtendalia miyabei* Matsumura]	*R.semialata*	Formosa	Patch 1938
[*Schlechtendalia sinensis* Walker]	*R.semialata*	?China	Patch 1938
[*Melaphis peitan* Tsai & Tang]	*R.javanica*	China	Tsai and Tang 1946
Schlechtendalia elongallis Tsai & Tang (*Meitanaphis*)	*R.punjabensis sinica*	China	Tsai and Tang 1946
Slavum esfandiarii Davatchi & Remaudiere	*P.mutica*	Iran	Davatchi 1958
Slavum lentiscoides Mordvilko	*P.mutica*	Iran	Davatchi 1958
	P.vera	USSR	Mordvilko 1927
Slavum mordvilkoi Kreutzberg	*P.vera*	Iran	Davatchi 1958
Slavum wertheimae H. Ris Lambers	*P.mutica, P.vera, P.atlantica*	Iran, Israel	Davatchi 1958
Smynthurodes betae Westwood (P)	*P.mutica*	Iran	Davatchi 1958
	P.atlantica	Algeria	Davatchi 1958
	P.vera	Syria	Davatchi 1958
[*Trifidaphis phaseoli* Passerini]	*P.terebinthus*	France	Patch 1938
	P.mutica	France	Patch 1938
	P.atlantica	Israel	Koach and Wool 1977
Therioaphis langloisi Remaudiere & Leclant	*P.lentiscus*	France	Richards 1976
Toxoptera aurantii Boyer de Fonscolombe	*P.atlantica*	Morocco	Davatchi 1958
Toxoptera odinae van der Goot (P) (*Aphis*)	*R.semialata*	Korea	Patch 1938
	R.javanica	Formosa	Patch 1938

Taxa	*Hosts*	*Locality*	*Reference*
[*Arimakia araliae* Matsumura]	*R.vernicifera*	Japan	Patch 1938
Xerophylophis nevskyi (Kreutzberg) (*Brachyunguis*)	*P.vera*	USSR, Iran	Davatchi 1958
CERCOPIDAE			
Lepyronia quadrangularis (Say) (P)	*R.glabra*	N.Amer.	Doering 1942
CICADELLIDAE			
Bolinlila typhlocyboides Signoret	*R.vulgaris*	Uganda	LePelley 1959
Empoasca chelata DeLong & Davidson (O)	*R.copallina*	VA	Poos and Wheeler 1943
Empoasca fabae (Harris) (P)	*R.copallina, R.glabra, R.typhina*	USA	Poos and Wheeler 1943
Erythroneura maculata (Gillette)	*R.*sp.	Ontario	Beirne 1956
Idiocerus alkani Wagner	*P.vera*	Turkey	Gentry 1965
Idiocerus stali Fieber	*P.vera*	Turkey, Syria, Iran	Gentry 1965
Platymoideus trilineatus Ball	*R.trilobata*	CA	Ball 1932
CLASTROPTERIDAE			
Clastroptera obtusa (Say)	*R.radicans*	MI	Gillis 1971
COCCIDAE			
Anapulvinaria pistaciae Bodenheimer	*P.palaestina*	Israel	Davatchi 1958
	P.vera	Iran, Syria	Davatchi 1958
	P.khinjuk	Iran	Davatchi 1958
Ceroplastes rusci (Linn.)	*P.vera*	Iran	Davatchi 1958
	P.vera	Aegean	Ulu et al. 1974
	P.lentiscus	Italy, n.Afr.	Davatchi 1958
	Schinus terebinthifolius	Israel	Davatchi 1958
Ceroplastes sinensis Del Guerico	*P.lentiscus*	France	Davatchi 1958
Ceroplastes toddaliae Hall (P)	*R.*sp.	Rhodesia	Hall 1931
Eulecanium rugulosum (Archangelskaya)	*P.vera, P.khinjuk*	Iran	Davatchi 1958
Filippia oleae Costa	*P.lentiscus*	Italy, n.Afr.	Davatchi 1958
Filippia oleae var. *rhusae* Hall	*R.*sp.	Rhodesia	Hall 1932
Floraphis chovi Xiang	*R.potanini*	China	Xiang 1980
Howardia rhusae Hall	*R.villosa*	Rhodesia	Hall 1928
Kaburagia rhusicola Takagi	*R.potanini*	China	Xiang 1980
Kaburagia ovatirhusicola Xiang	*R.potanini*	China	Xiang 1980
Lecanium corni Bouche	*P.vera*	USSR	Davatchi 1958
Saissetia coffeae (Walker) (P)	*R.*sp., *Schinus* sp.	?CA–FL	Johnson and Lyon 1976
Saissetia oleae (Bernard) (P)	*P.lentiscus*	Italy	Davatchi 1958
	P.vera	Aegean	Ulu et al. 1974
	*R.*sp., *Schinus* sp.	?CA	Johnson and Lyon 1976

Taxa	*Hosts*	*Locality*	*Reference*
Salicicola vayssierei (Balachowsky)	*R.pentaphylla*	Morocco	Balachowsky 1928
	R.tripartita	Mauritania	Balachowsky 1958
DIASPIDIDAE			
Abgrallaspis townsendi (Cockerell)	*R.*sp.	FL	Dekle 1976
Andaspis mori Ferris (P)	*P.*sp.	?	Borchsenius 1966
Aonidia badia Brain	*R.*sp.	S.Afr.	Balachowsky 1958
Aonidia marginalis Brain	*R.*sp.	S.Afr.	Balachowsky 1958
Aonidia rhusae Brain	*R.*sp.	S.Afr.	Balachowsky 1958
Aonidiella aurantii (Maskell) (P)	*R.*sp., *P.*sp., *Schinus*	?	Borchsenius 1966
Aspidaspis braunschvigi (Rungs) (O)	*P.atlantica*	CA	McKenzie 1956
Aspidiotus heredae Vallot (P)	*R.*sp.	w.USA	
Aspidiotus nerii Bouche (P)	*P.atlantica, P.lentiscus, P.palaestinae*	Israel	Gerson (pers. comm.)
	*P.*sp., *R.*sp., *Schinus*	?	Borchsenius 1966
Bantudiaspis loranthi (Hall) (P)	*R.*sp.	?	Borchsenius 1966
Chrysomphalus aonidum (Linn.) (P)	*R.*sp.	?	Borchsenius 1966
Chrysomphalus bifasciculatus Ferris (P)	*R.*sp.	CA	McKenzie 1956
Chrysomphalus dictyospermi (Morgan) (P)	*R.*sp., *P.*sp., *Schinus*	?	Borchsenius 1966
Clavaspis covilleae (Ferris) (P)	*R.trilobata*	CA	Kimball 1965
Diaspidiotus ancyclus (Putnam)	*R.*sp.	FL	Dekle 1976
Diaspidiotus pectinatus (Lindinger) (P)	*R.*sp., *Schinus*	?	Borchsenius 1966
Diaspis carmanicus (Davatchi & Balachowsky)	*P.khinjuk*	Iran	Davatchi 1958
Diaspis syriaca Lindinger	*P.khinjuk*	Iran	Davatchi 1958
	P.terebinthus	Greece, Turkey, Palestine	Davatchi 1958
	P.vera	Syria	Davatchi 1958
	P.lentiscus, P.palaestina, P.atlantica	Israel	Gerson (pers. comm.)
Duplaspidiotus claviger (Cockerell)	*R.radicans*	FL	Dekle 1976
Dynaspidiotus britannicus (Newstead) (P)	*P.*sp.	?	Borchsenius 1966
	P.lentiscus	Israel	Gerson (pers. comm.)
Dynaspidiotus ericarum (Goux) (P)	*P.*sp.	?	Borchsenius 1966

Taxa	Hosts	Locality	Reference
Epidiaspis gennadii (Leonardi)	*P.vera, P.lentiscus, P.terebinthus*	Greece, Turkey, Yugo., Italy, Cyprus, Syria	Davatchi 1958
	P.vera	Iran	Davatchi 1958
	P.atlantica, P.lentiscus, P.palaestina	Israel	Gerson (pers. comm.)
Epidiaspis leperii Signoret	*P.vera*	Greece	Davatchi 1958
Fiorina fiorinae (Targioni) (P)	*R.*sp., *P.*sp.	CA	McKenzie 1956
Furchadiaspis zamiae (Morgan) (P)	*R.*sp.	?	Borchsenius 1966
Gymnaspis faurei Brain	*R.*sp.	?	Borchsenius 1966
Hemiberlesia lataniae (Signoret) (P)	*R.*sp., *P.*sp., *Schinus*	?	Borchsenius 1966
Hemiberlesia rapax (Comstock) (P)	*R.*sp., *Schinus*	?	Borchsenius 1966
Hemiberlesia 3 spp.	*S. molle*	Tanz.	Bohlem 1973
Inchoaspis dentilobis Newstead (P)	*R.natalensis*	Uganda	LePelley 1959
Lepidosaphes fiorii Leonardi	*R.*sp.	?	Borchsenius 1966
Lindingaspis opima Silvestri	*R.natalensis*	Kenya	Bohlem 1973
Lindingaspis rossi (Maskell) (P)	*R.*sp., *Schinus*	?	Borchsenius 1966
Meitanaphis microgallis Xiang	*R. potanini*	China	Xiang 1980
Melanaspis corticosa (Brain) (P)	*R.*sp., *Schinus*	?	Borchsenius 1966
Melanaspis inopinatus Leonardi (P)	*P.vera, P.mutica*	Iran	Davatchi 1958
	P.vera	Syria	Davatchi 1958
	P.vera	Aegean	Ulu et al. 1974
	P.atlantica, P.palaestinae	Israel	Gerson (pers. comm.)
Mytilaspis conchiformis (Gmelin) (P)	*P.*sp.	?	Borchsenius 1966
Neopinnaspsis harperi McKenzie (P)	*Schinus molle, S.terebinthifolia*	CA	McKenzie 1956
Parlatoria oleae (Colvee) (P)	*P.vera*	Iran	Davatchi 1958
	P.vera	Aegean	Ulu et al. 1974
	P.atlantica	Israel	Gerson (pers. comm.)
Parlatoria pergandii (Comstock) (P)	*P.lentiscus*	Israel	Gerson (pers. comm.)
Parlatoria yunnanensis Ferris (P)	*P.*sp.	?	Borchsenius 1966
Parlatoreopsis chinensis (Marlott) (P)	*P.*sp., *R.*sp.	CA	McKenzie 1956
Pinnaspis strachani (Cooley)	*R.leucantha*	FL	Dekle 1976
Pistaciaspis (=*Lepidosaphes*) *pistaciae* (Archangelskaya) (incl. f. *pistacicola* Borchsenius)	*P.*spp.	c.Asia, Iraq, Turkey, Syria	Davatchi 1958
	P.vera	Aegean	Ulu et al. 1974

Taxa	*Hosts*	*Locality*	*Reference*
	P.vera, P.mutica, P.khinjuk	Iran	Davatchi 1958
	P.atlantica, P.palaestinae	Israel	Gerson (pers. comm.)
Pseudaonidia duplex (Cockerell) (P)	*R.*sp.	?	Borchsenius 1966
	*R.toxicodendron, R.*sp.	FL	Dekle 1976
Pseudaulacaspis pentagona (Targioni)	*R.*sp.	FL	Dekle 1976
Pulvinaria innumerabilis (Rathvon) (P)	*R.*sp.	VA	Williams and Kostarab 1972
Pulvinaria maclurae Kenn. (P)	*R.*sp.	N.Amer.	Felt 1906
Quadraspidiotus braunschvigi Rungs	*P.atlantica*	Morocco	Davatchi 1958
Quadraspidiotus juglansregiae (Comstock)	*R.*sp.	FL	Dekle 1976
Quadraspidiotus labiatarum (Marchal) (P)	*R.*sp.	?	Borchsenius 1966
Quadraspidiotus lentiscularis Lindinger (P)	*P.lentiscus*	Yugo., Italy	Davatchi 1958
Quadraspidiotus perniciosus (Comstock) (P)	*R.*sp., *Schinus*	?	Borchsenius 1966
Rolaspis lounsburyi (Cooley)	*R.*sp.	?	Borchsenius 1966
Salicicola davatchii Balachowsky & Kaussari	*P.khinjuk*	Iran	Davatchi 1958
	P.vera	Turkey	Davatchi 1958
Salicicola pistaciae Lindinger	*P.*spp.	c.Asia, Iraq, Greece, Turkey, Israel	Davatchi 1958
	P.vera	Syria	Hammad and Mohamed 1965
	P.mutica, P.khinjuk	Iran	Davatchi 1958
	P.lentiscus	Cyprus	Davatchi 1958
Salicicola vayssieri Balachowsky (M)	*R.*sp.	Sahara, sub-Sahara	Davatchi 1958
Selenaspidus sp.	*S.molle*	Tanz.	Bohlem 1973
Suturaspis pistaciae (Ling.)	*P.atlantica, P.palaestinae*	Israel	Gerson (pers. comm.)
Tecaspis mytilaspiformis (Newstead)	*R.*sp.	?	Borchsenius 1966
JASSIDAE			
Idiocerus brusinae Horvath	*P.vera*	Aegean	Ulu et al. 1974
Idiocerus pistaciae Anagnostopoulos	*P.vera*	Greece	Davatchi 1958
Idiocerus stali Fieber	*P.vera, P.mutica, P.khinjuk*	Iran	Davatchi 1958
	P.vera	Aegean	Ulu et al. 1974
	P.vera	USSR	Romanenko 1972
	P.vera	Syria	Schneider 1958
PSEUDOCOCCIDAE			
Heliococcus stachyus (Ehrhorn) (P)	*R.diversiloba*	CA	McKenzie 1967

Taxa	*Hosts*	*Locality*	*Reference*
Paracoccus sp.	*R.coriaria*	Israel	Gerson (pers. comm.)
Pseudococcus citri Risso (P)	*R.*sp.	Kenya	James 1933
Pseudococcus longisetosus Ferris (P)	*R.diversiloba*	CA	McKenzie 1967
Pseudococcus maritimus (Ehrhorn) (P)	*R.diversiloba*	USA, Can., Mex.	McKenzie 1967
PSYLLIDAE			
Agonoscena menozzii Laing.	*P.lentiscus*	Italy	Davatchi 1958
Agonoscena sp. near *menozzii*	*P.khinjuk*	Iran	Davatchi 1958
Agonoscena succinta DeGeer (O)	*P.vera, P.lentiscus, P.mutica*	USSR, Italy	Davatchi 1958
Agonoscena targionii Lichtenstein	*P.vera, P.mutica, P.khinjuk, P.terebinthus, P.lentiscus*	Iran	Davatchi 1958
	*P.*spp.	Med.Eur., n.Afr.	Davatchi 1958
	P.vera	Syria	Schneider 1958
	P.vera	USSR	Romanenko 1972
Arytania capensis Pettey	*R.*sp.	S.Afr.	Pettey 1933
Arytania karrooensis Pettey	*R.*sp.	S.Afr.	Pettey 1933
Arytania ribesiae Crawford (P)	*R.trilobata*	CO	Crawford 1914
Calophya californica Schwarz	*R.ovata*	CA	Johnson and Lyon 1976
	R.integrifolia	CA	Crawford 1914
Calophya flavida Schwarz	*R.glabra*	Wash., D.C.	Crawford 1914
Calophya nigripennis Riley	*R.copalina*	Wash., D.C.	Crawford 1914
	R.glabra	s.USA	Johnson and Lyon 1976
Calophya rhois F. Loew	*C.coggygria*	Eur.	Houard 1909
	*R.*sp.	USA	Crawford 1914
Calophya triozomima Schwarz	*R.trilobata*	sw.USA	Johnson and Lyon 1976
	R.aromatica	MO	Johnson and Lyon 1976
Psylla sp.	*P.vera*	Iran	Davatchi 1958
	P.vera	Aegean	Ulu et al. 1974
Trioza mexicana Crawford	*R.*sp.	Mex.	Crawford 1914
HYMENOPTERA			
ANDRENIDAE (POLLINATORS)			
Andrena 11 spp.	*R.aromatica*	IL	Robertson 1896
ANTHOPHORIDAE (NEST BUILDERS)			
Ceratina calcarata Robertson	*R.typhina, R.glabra*	VA	Krombein 1960
Ceratina metallica H. S. Smith	*R.typhina, R.glabra*	VA	Krombein 1960
APIDAE (POLLINATORS)			
Apis mellifera Linn.	*R.glabra*	NY	Hayes 1975

Taxa	*Hosts*	*Locality*	*Reference*
		LA	Kauffeld et al. 1975
ARGIDAE			
Arge coccinea (Fab.)	*R.*spp.	e.N.Amer.	Krombein et al. 1979
Arge humeralis (Beauvois)	*R.radicans*	e.N.Amer.	Regas-Williams and Habeck 1979
CHALCIDIDAE (PARASITES)			
Brachymeria intermedia Nees (ex *Ocneria terebynthina*)	*P.vera, P.mutica*	Iran	Davatchi 1958
Euchalcidia nigripes Fonscolombe (ex *Recurvaria pistaciicola*)	*P.vera, P.mutica*	Iran	Davatchi 1958
COLLETIDAE (POLLINATORS)			
Colletes sp.	*R.aromatica*	IL	Robertson 1896
ENCYRTIDAE (PARASITES)			
Copidosoma sp. (ex *R. pistaciicola*)	*P.vera, P.mutica*	Iran	Davatchi 1958
Psyllaeophagus sp. (ex *Agonoscena targionii*)	*P.lentiscus*	Iran	Davatchi 1958
EUMENIDAE (NEST BUILDERS)			
Leptochilus ornatus Saussure	*R.typhina, R.glabra*	VA	Krombein 1960
EUPELMIDAE (PARASITES)			
Eupelmus martellii Masi (ex *Megastimus pistaciae*)	*P.terebinthus*	France	Davatchi 1958
Eupelmus urozonus Dalman (ex *M. pistaciae*)	*P.vera*	Tunisia	Davatchi 1958
EURYTOMIDAE (PARASITES & SEED-FEEDERS)			
Bruchophagus setigerus Mayr (ex *M. pistaciae*)	*P.*sp.	USSR	Davatchi 1958
Decatoma biguttata Swederus (ex *M. pistaciae, Cyrtoptyx*)	*P.terebinthus*	Iran	Davatchi 1958
Eurytoma compressa Bugbee	*R.trilobata, R.virens, R.ovata?*	CA, Mex.	Bugbee 1975
Eurytoma pistaciae Rondani	*P.*sp.	Greece, Italy	Davatchi 1958
Eurytoma plotnikovi Nikolskaya	*P.vera*	Iran	Davatchi 1958
	P.vera	USSR	Romanenko 1972
	P.vera	Syria	Schneider 1958
Eurytoma rhois Crosby complex	*R.hirta, R.copallina, R.glabra, R.typhina*	NC, e.N.Amer.	Neunzig 1964
Eurytoma rosae Nees (ex *M. pistaciae*)	*P.terebinthus*	Sicily	Davatchi 1958
Eurytoma seminis Bugbee	*R.trilobata*	KS, OK	Krombein et al. 1979
Eurytoma terebinthus Rondani	*P.terebinthus*	Italy	Davatchi 1958
eurytomid sp.	*R.integrifolia, R.ovata*	CA	Young 1972

Taxa	*Hosts*	*Locality*	*Reference*
HALICTIDAE (POLLINATORS)			
Agapostemon sp.	*R.aromatica*	IL	Robertson 1896
Augochlora sp.	*R.aromatica*	IL	Robertson 1896
Augochloropsis metallica (Fab.) (*A. viridissimus* Viereck)	*R.glabra*	CT	Rohwer 1916
Halictus 7 spp.	*R.aromatica*	IL	Robertson 1896
ICHNEUMONIDAE (POLLINATORS)			
sp.	*R.aromatica*	IL	Robertson 1896
MELITTIDAE (POLLINATORS)			
Macropsis patellata Patton	*R.glabra*	CT	Rohwer 1916
PTEROMALIDAE (PARASITES)			
Cyrtoptyx pistaciae Nikolskaya (ex *M.pistaciae*)	*P.vera, P.mutica*	Iran	Davatchi 1958
	P.terebintus	France	Davatchi 1958
	P.atlantica	Morocco	Davatchi 1958
Pachyneuron sp. (ex aphid galls)	*P.atlantica*	Morocco	Davatchi 1958
SPHECIDAE (NEST BUILDERS/ PREDATORS)			
Ectemnius stirpicola (Packard)	*R.typhina, R.glabra*	VA	Krombein 1960
Passaloecus annulatus (Say)	*R.glabra*	VA	Krombein 1960
Pemphredon harbecki Rohwer	*R.typhina, R.glabra*	VA	Krombein 1960
Pemphredon lethifer (Schuckard)	*R.glabra*	VA	Krombein 1960
Stigmus americanus Packard	*R.*sp.	CT	Rohwer 1916
TENTHREDINIDAE (NEST BUILDERS)			
Hemitaxonus dubitatus (Norton)	*R.typhina*	VA	Krombein 1960
Hylotoma humeralis Beauvois	*R.radicans*	CT	MacGillivray 1916
TORYMIDAE			
Idiomacromerus bimaculipennis Crawford	*R.glabra*	KS	Lovell 1964
Megastigmus pistaciae Walker	*P.vera, P.mutica*	Iran	Davatchi 1958
	P.vera	USSR	Romanenko 1972
	P.terebinthus	Italy, France	Davatchi 1958
	P.atlantica	Morocco	Davatchi 1958
torymid sp.	*R.integrifolia, R.ovata*	CA	Young 1972
VESPIDAE (NEST BUILDERS)			
Stenodynerus perennis (Saussure)	*R.typhina*	VA	Krombein 1960
XYLOCOPIDAE (POLLINATORS)			
2 spp.	*R.aromatica*	IL	Robertson 1896

Taxa	*Hosts*	*Locality*	*Reference*
LEPIDOPTERA			
ARCTIIDAE			
Diacrisia virginica (Fab.) (P)	*R.*sp.	N.Amer.	Tietz 1972
Halisodota caryae (Harris) (P)	*R.*sp.	N.Amer.	Tietz 1972
	R.typhina	Ontario	McGugan 1958
Halisodota maculata (Harris) (P)	*R.toxicodendron, R.vernix*	N.Amer.	Tietz 1972
Hypantria cunea (Drury) (P)	*Cotinus coggygria*	N.Amer.	Tietz 1972
arctiid sp.	*S. molle*	Tanz.	Bohlem 1973
EUCOSMIDAE			
Eucosma melanoneura Meyrick	*R.semialata*	India	Mathur and Singh 1960
GELECHIIDAE			
Anacampis rhoifructella Clemens	*R.typhina*	NY	Forbes 1923
Aristotelia rhoisella Busck	*R.*sp.	CA	Busck 1934
Gelechia walsinghami Dietz	*R.typhina*	NY	Forbes 1923
Holcocera chalcofrontella Clemens	*R.*sp.	NY	Forbes 1923
Schneidereria (Recurvaria) pistaciella Weber	*P.vera*	Syria	Schneider 1958
Schneidereria pistaciicola Danilevsky	*P.vera, P.mutica*	Iran	Davatchi 1958
	P.vera	USSR	Romanenko 1972
Stenovalva disclusa (Meyrick)	*R.tripartita*	Israel	Sattler 1982 (ex D. Furth, *in litt.*)
Teleiodes decorella (Haworth) (P)	*C. coggygria*	Eur.	Davatchi 1958
	*R.*sp.	?Eur.	Sattler 1982
Teleiopsis sarcochroma (Walsingham)	*R. tripartita*	Israel, Jordan, S.Yemen, n.Afr., Canary Is.	Sattler 1982
Telphusa humeralis Zeller (P)	*P.*sp., *Rhus*	USSR	Davatchi 1958
	P.terebinthus	France	Davatchi 1958
Telphusa modesta (Danilevsky)	*P.vera*	USSR	Romanenko 1972
Thyrsostoma querini (Stainton)	*P.terebinthus*	France	Davatchi 1958
	P.atlantica	Morocco	Davatchi 1958
GEOMETRIDAE			
Amphidasis cognataria Guenee (P)	*R.typhina*	N.Amer.	Teitz 1972
Antepione arcasaria (Walker) (O)	*R.copallina, R.glabra, R.typhina*	N.Amer.	Tietz 1972
Cemiostoma sp.	*P.vera, P.mutica, P.khinjuk?*	Iran	Davatchi 1958
Cingilia catenaria (Drury) (P)	*R.toxicodendron, R.*sp.	N.Amer.	Tietz 1972

Taxa	*Hosts*	*Locality*	*Reference*
Dichorda iridaria (Guenee) (M)	*R.*sp.	FL	Forbes 1948
Ellopia somniaria Hulst (P)	*R.toxicodendron*	N.Amer.	Tietz 1972
Eupithecia dubiosa Dietze	*P.terebinthus*	Cyprus, Lebanon	Davatchi 1958
	P.vera	Lebanon	Gentry 1965
Melanolophia canadaria (Guenee) (P)	*R.typhina*	N.Amer.	Tietz 1972
Nemoria bifilata (Walker)	*R.copallina*	SC	Ferguson 1975
Oulobophora externata (Herrich-Schaeffer)	*P.mutica*	Iran	Davatchi 1958
	*P.*sp.	Iraq	Wiltshire 1957
	*P.*sp.	Lebanon	Davatchi 1958
	P.vera	Lebanon	Gentry 1965
Rhoptria marginata Staudinger	*P.*sp.	Iran	Davatchi 1958
Sabulodes caberata Guenee (P)	*R.*sp.	N.Amer.	Tietz 1972
Sabulodes thisoaria Guenee (P)	*R.*sp.	NY	Forbes 1948
Tetracis crocallata Guenee (P)	*R.*sp.	NY	Forbes 1948
Zamacra flabellaria Heeger. (P)	*P.vera*	Iran	Davatchi 1958
Gracillariidae			
Acrocercops deversa Meyrick	*R.semialata*	India	Fletcher 1933
Caloptilia diversilobiella Opler	*R.diversiloba*	s.CA	Opler 1969
Caloptilia ovatiella Opler	*R.ovata, R.laurina, R.integrifolia*	s.CA	Opler 1969
Caloptilia (=Gracilaria) rhoifoliella (Chambers)	*R.radicans, R.copallina*	e.USA	Forbes 1923
	R.toxicodendron	Manitoba	Criddle, 1927
Caloptilia rhusina Vari	*R.lucida*	S.Afr.	Vari 1961
Caloptilia schinella (Walsingham) (O)	*S.molle, P.*spp.	?	Sattler 1982
Corethrovalva paraplesia Vari	*R.*sp.	S.Afr.	Vari 1961
Lithocolletis guttifinitella (Clemens)	*R.toxicodendron*	NY	Forbes 1923
	R.toxicodendron	Manitoba	Criddle 1927
gracillariid 2 spp.	*S.molle*	Tanz.	Bohlem 1973
Hesperiidae			
Acleros mackenii Trimen (O)	*R.coriacius*	e.Afr.	Van Someren 1974
Eagris leucetia Hewitson	*R.villosa, R.vulgaris*	e.Afr.	Van Someren 1974
Lasiocampidae			
Bombycomorpha nupta Felder (=*bifascia* Walker)	*R.*sp., *Schinus molle*	S.Afr.	Brain 1929
Bombycomorpha pallida Distant	*R.*sp., *Schinus molle*	S.Afr.	Brain 1929
Lebeda nobilis Walker	*R.parviflora*	India	Mathur and Singh 1960

Taxa	Hosts	Locality	Reference
Lechriolepis basirufa Strand (P)	*R.vulgaris*	e.Afr.	Austara and Jones 1971
Lechriolepis nigrivenis Strand	*R.incana*	e.Afr.	Jackson 1932
Malacosoma frugile (Stretch)	*R.trilobata*	NM	J. M. Schmid (pers. comm.)
Pachypasa otus (Drury) (P)	*P.mutica*	Iran	Davatchi 1958
	*P.*sp.	Iraq	Wiltshire 1957
	P.vera	Syria	Schneider 1958
Schausinna affinis Aurivillius (P)	*R.*sp.	e.Afr.	Austara and Jones 1971
Trichiura pistaciae Wiltshire	*P.mutica*	Iran	Davatchi 1958
LYCAENIDAE			
Alaena subrubra Bethune-Baker	*R.vulgaris*	Uganda	LePelley 1959
Anthene definita Butler (P)	*R.vulgaris*	Uganda	LePelley 1959
	R.incana	e.Afr.	Van Someren 1974
Calycopis cecrops (Fab.) (P)	*R.*sp.	TX	D. Harvey (via L. and J. Miller) (pers. comm.)
Leptotes cassius theonus (Lucas)	*R. radicans*	FL	Kimball 1965
Pseudargiolus lycaenopsis (Boisduval & LeConte)	*R.*sp.	N.Amer.	Tietz 1972
Spindasis appeles nairobiensis Sharpe	*R.vulgaris*	e.Afr.	Van Someren 1974
LYMANTRIIDAE			
Euproctis fasciata Walker	*R.vulgaris*	Uganda	LePelley 1959
Euproctis kharghalica Moore	*P.*spp.	USSR, Iran	Davatchi 1958
Hemerocampa leucostigma (Abbot & Smith) (P)	*Cotinus coggygria*	N.Amer.	Tietz 1972
Lymantria dispar (Linn.) (P)	*R.glabra, R.toxicodendron, R.typhina*	N.Amer.	Tietz 1972
	P.vera	Aegean	Ulu et al. 1974
Lymantria modesta Walker	*R.vulgaris*	Uganda	LePelley 1959
Ocneria audeoudi Brandt	*P.mutica*	Iran	Davatchi 1958
Ocneria terebynthina Staudinger	*P.vera, P.mutica*	Iran	Davatchi 1958
	*P.*sp.	Iraq	Wiltshire 1957
Parocneria terebinti Ferriere	*P.terebintus*	USSR, Med.	Davatchi 1958
NEPTICULIDAE			
Stigmella (=*Nepticula*) Braun *intermedia* Braun	*R.*spp.	NY	Forbes 1923
Stigmella promissa (Staudinger)	*P.vera, P.terebinthus, P.mutica, P.lentiscus*	USSR, c.Asia, Med. Eur.	Davatchi 1958
	P.khinjuk	Iran	Davatchi 1958
	Cotinus coggygria, Rhus coriaria	USSR	Davatchi 1958
	*R.*sp.	?	Sattler 1982
Stigmella rhoifoliella Braun	*R.radicans*	NY	Forbes 1923
Weberina lentiscella Groschke	*P.lentiscus*	Sicily	Groschke 1944

Taxa	*Hosts*	*Locality*	*Reference*
NOCTUIDAE			
Acronicta longa Guenee (P)	*R.toxicodendron*	N.Amer.	Tietz 1972
Actebia fennica (Tauscher) (P)	*R.*sp.	N.Amer.	Tietz 1972
Agrotinae 2 spp.	*S.molle*	Tanz.	Bohlem 1973
Cacoecia argyrospila Walker	*R.toxicodendron*	Manitoba	Criddle 1927
Epipaschia zelleri Grote	*R.toxicodendron*	Manitoba	Criddle 1927
Eutelia adoratrix Staudinger (M)	*P.mutica*	Iran	Davatchi 1958
	P.lentiscus, P.terebinthus, Cotinus coggygria	France	Davatchi 1958
	*P.*sp.	Iraq	Wiltshire 1957
Eutelia adulatrix Hubner (M)	*P.*sp.	USSR, Lebanon, Med.	Davatchi 1958
	*P.*sp.	Iraq	Wiltshire 1957
	*Cotinus coggygria, P.*sp.	?India	Sevastopulo 1941
Eutelia amatrix Hubner (M)	*R.vulgaris*	Uganda	LePelley 1959
Eutelia pulcherrima (Grote) (M)	*R.vernix, R.*sp.	NY, FL	Forbes 1954
	R.toxicodendron	e.USA	D. Schweitzer (pers. comm.)
Marathyssa basalis Walker (M)	*R.radicans*	NY, FL	Forbes 1954, Kimball 1965
Marathyssa inficita Walker (M)	*R.typhina*	NY, FL	Forbes 1954, Kimball 1965
	R.copallina	e.USA	D. Schweitzer (pers. comm.)
	R.glabra	Brit.Col.	Leech 1943
Ophiusa (=Anua) trihaca (Cramer)	*P.terebinthus*	Lebanon	Davatchi 1958
	P.mutica, P.khinjuk	Iran	Davatchi 1958
	P.terebinthus, P.lentiscus, Rhus. sp.	France	Davatchi 1958
	*P.*sp.	Iraq	Wiltshire 1957
	R.vulgaris	Uganda	LePelley 1959
	S.molle	Tanz.	Bohlem 1973
	R.coriaria, P.lentiscus	?India	Sevastopulo 1941
Paectes oculatrix (Guenee) (M)	*R.toxicodendron*	Manitoba	Criddle 1927
	R.radicans	NY, FL	Forbes 1954
Peridroma margaritosa (Haworth) (P)	*R.copallina, R.*sp.	N.Amer.	Tietz 1972
Phlegetonia catephiodes Guenee	*R.vulgaris*	Uganda	LePelley 1959
Polia cristifera (Walker) (P)	*R.*sp.	N.Amer.	Tietz 1972
Polia lutra (Guenee) (P)	*R.*sp.	N.Amer.	Tietz 1972
Pyrrhia umbra (Hufangel) (P)	*R.*sp.	N.Amer.	Tietz 1972
	R.typhina	e.Can.	Prentice 1962
Selepa celtis Moore	*R.semialata*	India	Mathur and Singh 1960

Taxa	*Hosts*	*Locality*	*Reference*
NOTODONTIDAE			
Datana major Grote & Robinson	*R.glabra*?	N.Amer.	Tietz 1972
Datana ministra (Drury) (P)	*R.*sp.	N.Amer.	Tietz 1972
Datana perspicua Grote & Robinson (M)	*R.typhina, R.copallina, R.aromatica, R.*sp.	e.N.Amer.	Forbes 1948, Tietz 1972
	R.typhina	Ontario	Prentice 1962
Datana robusta Strecker (O)	*R.*sp.	N.Amer.	Tietz 1972
NYMPHALIDAE			
Charaxes varanes (Cramer)	*R.viminalis*	S.Afr.	Van Son 1979
Euphaedra uganda uganda Aurivillius	*R.subcoriaceus*	Uganda	Van Someren and Rogers 1935
PYRALIDAE			
Homoeosoma stypticellum Ragonot	*R.glabra*	NY	Forbes 1923
Laodamia combustella Herrich-Schaeffer (predatory on gall aphids)	*P.terebinthus, P.lentiscus*	USSR, Med.Eur.	Davatchi 1958
Meroptera provella Grote (O)	*C.coggygria*	USA	Forbes 1923
Moodna ostrinella (Clemens) (P)	*R.*spp.	NY, FL	Forbes 1923
Nephopteryx subfuscella (Ragonot)	*R.*sp.	FL	Kimball 1965
Salebria semiobscurella Hulst	*R.*sp.	NY	Forbes 1923
Schreckensteinia erythriella Clemens	*R.*spp.	NY	Forbes 1923
Sylepta balteata Fab.	*P.terebinthus, Cotinus*	USSR, Med.	Davatchi 1958
	R.vulgaris	Uganda	LePelley 1959
PSYCHIDAE			
Thyridopteryx ephemeraeformis (Haworth) (P)	*R.*spp., *Schinus terebinthifolius, S.molle*	e.N.Amer.	Davis 1964
SATURNIIDAE			
Bombycomorpha pallida Distant	*R.*sp., *Schinus molle*	S.Afr.	Brain 1929
Bombycomorpha bifascia Walker	*R.*sp., *Schinus molle*	S.Afr.	Brain 1929
Cirina forda (Westwood) (P)	*S.molle*	trop.Afr.	Pinhey 1972
	R.longifolia	trop.Afr.	Pinhey 1972
Citheronia regalis (Fab.) (P)	*R.copallina*	N.Amer.	Tietz 1972
	R.glabra, R.typhina	USA	Worth et al. 1979
Eacles imperialis (Drury) (P)	*R.glabra*	N.Amer.	Tietz 1972
Gonimbrasia belina (Westwood) (P)	*R.lancia, R.longifolia*	trop.Afr.	Pinhey 1972
Holocerina smilax (Westwood) (O)	*R.laevigata, R.longifolia, R.villosa*	trop.Afr.	Pinhey 1972
Hyalophora cecropia (Linn.) (P)	*R.typhina*	Can.	McGugan 1958
Imbrasia cytherea (Fab.)	*R.angustifolia, R.crenata, R.tomentosa, R.lucida, R.longifolia*	S.Afr.	Pinhey 1972

Taxa	*Hosts*	*Locality*	*Reference*
Imbrasia wahlbergi (Boisduval) (P)	*S.*sp.	s./e.Afr.	Pinhey 1972
Lobobunaea angasana (Westwood) (P)	*R.longifolia*	trop.Afr.	Pinhey 1972
Samia walkeri (Felder & Felder) (P)	*R.*sp.	N.Amer.	Tietz 1972
Saturnia pyri Linn. (P)	*P.*spp.	Iran, Lebanon	Davatchi 1958
Usta terpsichore (Maassen & Weymer)	*S.molle*	s./e.Afr.	Pinhey 1972
saturniid 5 spp.	*S.molle*	Tanz.	Bohlem 1973
SPHINGIDAE			
Compsogene panopus Cramer	*R.wallichii*	India	Mathur and Singh 1960
Oxyambulyx ochracea Butler	*R.semialata*	India	Mathur and Singh 1960
Oxyambulyx serceipennis agana Jordan	*R.insignis*	India	Mathur and Singh 1960
THAUMETOPOEIDAE			
Thaumetopoea jordana (Staudinger)	*R.tripartita*	Israel, Jordan	Furth and Halperin 1980
Thaumetopoea cheela Moore	*C.coggygria*	Pak.	Furth and Halperin 1980
Thaumetopoea solitaria Freyer	*P.vera*	Iran	Davatchi 1958
	P.vera, P.?spp.	Iraq	Wiltshire 1957
	P.vera	Aegean	Ulu et al. 1974
	P.terebinthus	Med.	Davatchi 1958
	P.atlantica, P.lentiscus, P.palaestina, P.vera, Schinus terebintifolius	Israel	Halperin 1983
Thaumetopoea solitaria iranica Ag.	*P.*sp.	Iran	Davatchi 1958
TINEIDAE			
Tinea pistaciae Anagnostopoulos	*P.vera*	Greece	Davatchi 1958
TORTRICIDAE			
Adoxophyes orana Fisch.v.Rssl. (P)	*P.lentiscus*	USSR, Eur.	Davatchi 1958
Amorbia humerosana Clemens	*R.*sp.	e.N.Amer.	Felt 1906
Archips purpuranus Clemens (P)	*R.*sp.	Can.	MacKay 1962
Archips rosaceana Harris (P)	*R.*sp.	N.Amer.	Felt 1906
Episimus argutanus (Clemens) (P)	*R.*sp.	NY, FL	Forbes 1923
	R.toxicodendron	Manitoba	Criddle 1927
	*R.toxicodendron, R.*sp.	Ontario, Quebec	MacKay 1959
Episimus augmentus (Zeller)	*Metopium(R.) toxiferum*	FL	Kimball 1965
Polychrosis rhoifructana Kearfott (O)	*R.*sp.	NY	Forbes 1923
	R.glabra	VA	MacKay 1959
tortricid sp.	*P.vera*	Aegean	Ulu et al. 1974

Taxa	*Hosts*	*Locality*	*Reference*
ORTHOPTERA			
ACRIDIDAE			
Lyrotyloides viridis Bey-Beinko (O)	*P.mutica*	Iran	Davatchi 1958
Schistocerca gregaria Forskal (P)	*P.vera*	Iran	Davatchi 1958
Schistocerca shoshone (Thomas) (P)	*R.trilobata*	AZ	Ball et al. 1942
Scudderia furcata furcata Brunner (P)	*R.typhina*	MI	Gangwere 1961
THYSANOPTERA			
PHLOEOTHRIPIDAE			
Liothrips jakhontovi Kreutzberg	*P.vera*	Iran	Davatchi 1958
THRIPIDAE			
Retrithrips syriacus Mayet (P)	*P.vera, P.terrebinthus, P.lentiscus*	Israel	Davatchi 1958
Rhynchothrips ilex (Moulton) (O)	*R.ovata*	CA	Bohart 1939
Scirtothrips citri (Moulton) (P)	*R.laurina*	CA	Tanigoshi and Nishu-Wong 1981
Taeniothrips inconsequens Uzel (P)	*P.vera*	USSR, Greece	Davatchi 1958
Taeniothrips meridionalis Priesner (P)	*P.spp.?*	USSR	Davatchi 1958
Thrips iranicus Yathintov	*P.vera*	Iran	Davatchi 1958
Thrips minutissimus Linn. form	*P.vera*	USSR	Davatchi 1958
Thrips pistaciae Yakhontov	*P.vera*	Iran	Davatchi 1958

KEY TO APPENDIX

? = *after* an arthropod or plant taxon, indicates some question in the determination; *before* a locality indicates uncertainty for that place.

Poison Ivy = *R.radicans* or *R.toxicodendron*, especially older records.

P = known to be polyphagous.

O = known to be oliphagous (= few other known hosts, usually in the same plant family).

M = known to be monophagous (usually host specific).

Localities are abbreviated: Med. = Mediterranean; N.Amer. = North America; Eur. = Europe; Afr. = Africa; all USA states are abbreviated according to the standard two letter postal code; e. = east; w. = west; s. = south; n. = north; c. = central.

pers. comm. = personal communications, courtesy of: Dr. U. Gerson (Rehovot, Israel); Dr. J. M. Schmid (Rocky Mountain Forest Experiment Station, Fort Collins, CO); Dr. D. F. Schweitzer (Boston, MA); Rev. J. Sullivan (St. Louis, MO); Drs. L. and J. Miller (Sarasota, FL).

APPENDIX REFERENCES

AUSTARA, O. and Jones, T. 1971. Host list and distribution of lepidopterous defoliators of exotic softwoods in East Africa. E. Afr. Agric. For. J. 36:401–413.

BAGDASARYAN, A. T. 1976. The eriophyid mites on subtropical plants in Armenia (Acari, Eriophyoidea). Zool. Sb. Erevan 17:210–222 (in Armenian with English summary).

BAKER, A. C. 1917. On the Chinese Gall (Aphididae, Homoptera). Ent. News 28(9):385.

BALACHOWSKY, A. 1928. Contribution à l'étude des coccides de l'Afrique mineure. Bull. Soc. Hist. Nat. Afr. N. 19(3):121–144.

BALACHOWSKY, A. 1956. Le cochenilles du continent Africain Noir. Vol. 1 Aspidiotini (part 1). Ann. Mus. Roy. Congo Belge Tervuren (N. S.) Sci. Zool. 3:1–142.

BALACHOWSKY, A. 1958. Le cochenilles du continent Africain Noir. Vol. 2 Aspidiotini (part 2), Odonoaspidini et Parlatorini. Ann. Mus. Roy. Congo Belge Tervuren (N. S.) Sci. Zool. 4:145–431.

BALL, E. D. 1932. The food plants of the leafhoppers formerly included in the genus *Platymetopius* Burm. Can. Ent. 64(11): 251–255.

BALL, E. D., Tinkham, E. R., Flock, R., and Vorhies, C. T. 1942. The grasshoppers and other Orthoptera of Arizona. Ariz. Agric. Exp. Sta., Tech. Bull. 93:255–373.

BARNES, H. F. 1948. Gall Midges of economic importance. Vol. 4: Gall Midges of ornamental plants and shrubs. London: Crosby Lockwood & Son.

BEIRNE, B. P. 1956. Leafhoppers (Homoptera: Cicadellidae) of Canada and Alaska. Can. Ent. Suppl. 2. 180 pp.

BLACKMAN, M. W. 1940. The scolytid beetles of the genus *Renocis* Casey, with descriptions of nine new species. Proc. U.S. Natl. Mus. 88(3084):373–401.

BLATCHLEY, W. S. 1926. Heteroptera or True Bugs of Eastern North America. Indianapolis: The Nature Publishing Co.

BLATCHLEY, W. S., and Leng, C. W. 1916. Rhynchophora or Weevils of North Eastern America. Indianapolis: The Nature Publishing Co.

BOHART, R. M. 1939. A new host plant and locality record for the Christmas Berry Thrips. J. Econ. Ent. 32(6):880–881.

BOHLEM, E. 1973. Crop pests in Tanzania and their control. Berlin: Verlag Paul Parey.

BOIVIN, G. and Stewart, R. K. 1982. The biology of five species of phytophagous mirids (Hemiptera: Heteroptera: Miridae) in apple orchard in southeastern Quebec. Ann. Soc. Ent. Quebec 27(3):148–151.

BORCHSENIUS, N. S. 1966. A catalogue of the Armoured Scale insects (Diaspidoidea) of the world. Moscow–Leningrad: Nauka.

BRAIN, C. K. 1929. Insect pests and their control in South Africa. Capetown: Die Nasionale Pers Beperk.

BRUCK, C. R. 1933. A new Bark Beetle from southern California. Can. Ent. 65(10):239–240.

BUGBEE, R. E. 1975. *Eurytoma* species from Mexico and Guatemala with synonyms and keys (Hymenoptera: Eurytomidae). Ann. Ent. Soc. Amer. 68(2):251–256.

BUSCK, A. 1934. Two new microlepidoptera from California. Bull. S. Calif. Acad. Sci. 33(2):74–78.

CHAMPLAIN, A. B. and Knull, J. N. 1922. Miscellaneous notes on Coleoptera. Can. Ent. 54:102–104.

CRAWFORD, D. L. 1914. A monograph of the jumping plant-lice or Psyllidae of U.S. Nat. Mus. Bull. 85:1–182.

CRIDDLE, N. 1927. Lepidoptera reared in Manitoba from Poison Ivy. Can. Ent. 59(5):99–101.

DAVIS, D. R. 1964. Bagworm moths of the Western Hemisphere. U.S. Nat. Mus. Bull. 244. 233 pp.

DAVATCHI, G. A. 1958. Etude biologique de la faune entomologique des *Pistacia* sauvages et cultivés. Rev. Pathol. végét. Ent. Agric. France. 37(1):3–166.

DEKLE, G. W. 1976. Florida Armoured Scale insects. Arthropods of Florida and neighboring land areas. Vol. 3. Gainesville: Florida Department of Agriculture.

DODGE, B. O. and Rickett, H. W. 1948. Diseases and pests of ornamental plants. New York: Ronald Press.

DOERING, K. C. 1942. Host plant records of Cercopidae in North America, north of Mexico (Homoptera). J. Kansas Ent. Soc. 15(2/3):65–92.

DRAKE, C. J. and Ruhoff, F. A. 1965. Lacebugs of the world: a catalog (Hemiptera: Tingidae). Bull. U.S. Nat. Mus. No. 243:1–634.

DUFFY, E. A. J. 1960. A monograph of the immature stages of Neotropical timber beetles (Cerambycidae). London: British Museum (Natural History). 327 pp.

DUFFY, E. A. J. 1968. A monograph of the immature stages of Oriental timber beetles (Cerambycidae). London: British Museum (Natural History). 434 pp.

EASTOP, V. F. and Hille Ris Lambers, D. 1976. Survey of the world aphids. The Hague: Dr. W. Junk b.v., Publishers.

EWING, H. E. 1917. Mites affecting the poison oak. Proc. Iowa Acad. 24:323–326.

FELT, E. P. 1905. Insects affecting park and woodland trees. N.Y. State Mus. Mem. 8(1):1–459.

FELT, E. P. 1906. Insects affecting park and woodland trees. N.Y. State Mus. Mem. 8(2):333–877.

FERGUSON, D. C. 1975. Host records for Lepidoptera reared in eastern North America. U.S. Dept. Agric., Tech. Bull. 152.

FLETCHER, T. B. 1933. Life histories of Indian Microlepidoptera (2nd Series), Cosmopterygidae to Neopseustidae. Imp. Council Agric. Res. (Dehli) Sci. Monogr. 4:1–85.

FORBES, W. T. M. 1923. Lepidoptera of New York and neighboring states. Primitive forms, Microlepidoptera, Pyraloids, Bombyces. Cornell Univ. Agric. Exp. Sta. Mem. No. 68.

FORBES, W. T. M. 1948. Lepidoptera of New York and neighboring states. Geometridae, Sphingidae, Notodontidae, Lymantriidae. Cornell Univ. Agric. Exp. Sta. Mem. No. 274.

FORBES, W. T. M. 1954. Lepidoptera of New York and neighboring states. Noctuidae. Cornell Univ. Agric. Exp. Sta. Mem. No. 329.

FROST, S. 1973. Hosts and eggs of *Blepharida dorothea* (Coleoptera: Chrysomelidae). Florida Ent. 56(2):120–122.

FURTH, D. G. 1982. *Blepharida* biology, as demonstrated by the Sacred Sumac Flea Beetle (*B. sacra* (Weise)). Spixiana, Suppl. 7:43–52.

FURTH, D. G. 1985. Some flea beetles and their food plants from Kenya (Chrysomelidae: Alticinae). Coleopterists Bull. 39(3):(in press).

FURTH, D. G. and Halperin, J. 1980. Observations on the phenology and biogeography of *Thaumetopoea jordana* (Staudinger) (Lepidoptera: Thaumetopoeidae). Israel J. Ent. 13:1–11.

Gangwere, S. K. 1961. A monograph of food selection in Orthoptera. Trans. Amer. Ent. Soc. 87(2/3):67–230.

Gentry, J. W. 1965. Crop insects of northeast Africa-southwest Asia. U.S. Dept. Agric. Tech. Bull. 273.

Gillis, W. T. 1971. The systematics and ecology of poison-ivy and the poison-oaks. Rhodora 73(793–796):71–540.

Groschke, F. 1944. Neues über Minerer aus dem Mittelmeergebiet. Mitt. Münchener Ent. Ges. 34(1):115–124.

Hall, W. J. 1928. Observations on the Coccidae of Southern Rhodesia. Bull. Ent. Res. 19(3):271–292.

Hall, W. J. 1931. Observations on the Coccidae of Southern Rhodesia. Trans. Ent. Soc. Lond. 79(2):285–303.

Hall, W. J. 1932. Observations on the Coccidae of Southern Rhodesia. Stylops 1(9):185–195.

Halperin, J. 1983. *Thaumetopoea solitaria* (Lepidoptera: Thaumetopoeidae) in Israel. Phytoparasitica 11(2):71–82.

Hammad, S. M. and Mohamed, M. H. 1965. Insect pests of *Pistacia* in the Aleppo District (Syria). Bull. Soc. Ent. Egypte 49:153–157.

Hayes, B. 1975. Nectar producing trees and the northeastern beekeeper. Amer. Bee J. 115(5):192.

Hellrigl, K. G. 1978. Ökologie und Brutpflanzen europäischer Prachtkäfer (Coleoptera, Buprestidae). Teil 1. Z. ang. Ent. 85:167–191, Teil 2. ibid. 85:253–275.

Hicks, S. D. 1955. A dermestid in Poison Ivy. Coleopt. Bull. 9(1):9.

Houard, C. 1909. Les Zoocecidies des plantes d'Europe et du Bassin de la Mediterranée. Vol. 2. Paris: Librairie Scientifique A. Herman et Fils.

Howden, H. F., Howden, A. T., and Ritcher, P. O. 1951. Insects feeding on Poison Oak (*Rhus toxicodendron*). Coleopterist Bull. 5(2):17–19.

Jackson, T. H. E. 1932. Life histories of some East African Lepidoptera. J. E. Afr. Uganda Nat. Hist. Soc. 42/43:173–176.

James, H. C. 1933. Taxonomic notes on the coffee mealybugs of Kenya Colony. Bull. Ent. Res. Lond. 24(3):429–436.

Janezic, F. 1978. *Dasyneura cotini* sp. n. (Diptera, Cecidomyiidae) in the leaf galls of *Cotinus coggygria* Scop. and *Contarina coronillae* sp. n. (Diptera, Cecidomyiidae) in the galls of *Coronilla emerus* L. and *C. emeroides* Boiss. & Spr. (in Slovenian with English summary). Biol. Vestnik (Ljubljana) 26(1):9–22.

Jeppson, L., Kiefer, H., and Baker, E. 1975. Mites injurious to economic plants. Los Angeles: University of California at Los Angeles Press.

JOHNSON, W. T. and Lyon, H. H. 1976. Insects that feed on trees and shrubs. Ithaca: Cornell University Press.

KAUFFELD, N. M., Hernandez, T., Wright, J., and Misaraca, S. 1975. Cucumber production in Louisiana with honeybees as pollinators. Part II, Nectar and pollen plants competing with cucumber for honeybee visits. Amer. Bee J. 115(4):156.

KEIFER, H. H. 1938. Eriophyid studies. Bull. Dept. Agric. Calif. 27:189–206, 301–323.

KEIFER, H. H. 1959. New Eriophyid mites. Ann. Ent. Soc. Amer. 52:649–657.

KEIFER, H. H. 1962. Eriophyid studies. Bull. Dept. Agric. Calif., Ser. B–6:1–20.

KEIFER, H. H. 1962a. Eriophyid studies. Bull. Dept. Agric. Calif., Ser. B–7:1–20.

KEIFER, H. H. 1966. Eriophyid studies. Bull. Dept. Agric. Calif., Ser. B–21:1–24.

KEIFER, H. H. 1979. Eriophyid studies. Bull. Dept. Agric. Calif., Ser. C(17):1–24.

KEIFER, H., Baker, E., Kono, T., Delfinado, M., and Styer, W. 1982. An illustrated guide to plant abnormalities caused by eriophyid mites in North America. Washington, D.C.: USDA, ARS, Agriculture Handbook No. 573.

KELTON, L. A. 1980. The insects and arachnids of Canada. Part 8. The Plant Bugs of the Prairie Provinces of Canada. Heteroptera: Miridae. Publ. 1703. 408 pp.

KIMBALL, C. 1965. Lepidoptera of Florida. Arthropods of Florida and neighboring land areas, No. 3. Gainesville: Florida Department of Agriculture.

KINGSLOVER, J. M., Batra, S. W. T., and J. A. Utmar 1984. A selected bibliography of insect-vascular plant associational studies. Washington, D.C.: USDA, ARS, Bibliogr. Liter. Agric. No. 27. 229 pp.

KIRKALDY, G. W. 1909. Catalogue of the Hemiptera (Heteroptera) with biological and anatomical references, lists of food-plants and parasites, etc. Berlin: F. L. Dames.

KNIGHT, H. 1941. The plant bugs, or Miridae of Illinois. Bull. Ill. Nat. Hist. Surv. 22:1–234.

KNOPF, H. E. 1971. Contributions to the knowledge of the insect fauna of the trees in Iraq, Part 1 Coleoptera. Z. ang. Ent. 69:82–87.

KNULL, J. N. 1930. Notes on Coleoptera, No. 2. Ent. News. 41:82–86, 101–102.

KNULL, J. N. 1946. The Long-Horned Beetles of Ohio (Coleoptera: Cerambycidae). Ohio Biol. Surv. Bull. 39:133–354.

KOACH, J. and Wool, D. 1977. Geographic distribution and host specificity of gall-forming aphids (Homoptera, Fordinae) on *Pistacia* trees in Israel. Marcellia 40:207–216.

KROMBEIN, K. 1960. Biological notes on some Hymenoptera that nest in Sumach pith. Ent. News 71(2):29–36, 63–69.

KROMBEIN, K. V., Hurd, P. D. Jr., Smith, D. R., and Burks, B. D. 1979. Catalog of the Hymenoptera in America north of Mexico, Symphyta and Aprocrita (Parasitica). Vol. 1. Washington, D.C.: Smithsonian Institution Press.

LEECH, H. B. 1943. *Marathyssa inficita* Walker at Oliver, British Columbia (Lepidoptera: Phalaenidae). Proc. Ent. Soc. Brit. Columbia 40:25.

LEPELLEY, R. H. 1959. Agricultural insects of East Africa. Nairobi: East African High Commission.

LINSLEY, E. G. 1962. The Cerambycidae of North America, Part 2. Univ. Calif. Publ. Ent. 19:1–102.

LINSLEY, E. G. 1963. The Cerambycidae of North America, Part 4. Univ. Calif. Publ. Ent. 21:1–165.

LOAN, C. C. and Matthews, R. 1973. A braconid parasite of a scolytid *Pityophthorus rhois* Swaine on *Rhus typhina*. Proc. Ent. Soc. Wash. 75:205–208.

LOVELL, J. F. 1964. An ecological study of *Rhus glabra*. Diss. Abst. 25(4):2190.

MACGILLIVRAY, A. D. 1916. In Viereck, H. L., The Hymenoptera, or wasp-like insects, of Connecticut. Conn. St. Geol.-Nat. Hist. Surv. Bull. 22.

MCGUGAN, B. M. 1958. Forest Lepidoptera of Canada. Vol. 1. Papilionidae to Arctiidae. Can. Agric., Sci. Serv., Div. For. Biol. Publ. 1034. 76 pp.

MACKAY, M. R. 1959. Larvae of the North American Olethrutidae (Lepidoptera). Can. Ent. Suppl. 10. 338 pp.

MACKAY, M. R. 1962. Larvae of the North American Tortricinae (Lepidoptera: Tortricidae). Can. Ent. Suppl. 28. 182 pp.

MCKENZIE, H. 1956. The armored scale insects of California. Bull. Calif. Insect Surv. No. 5.

MCKENZIE, H. 1967. Mealybugs of California. Los Angeles: University of California Press.

MATHUR, R. N. and Singh, B. 1960. A list of insect pests of forest plants of India and the adjacent countries. Indian For. Bull. (N.S.) No. 171 (7):1–60.

Monzen, K. 1929. Studies on some gall-producing aphids and their galls. Saito Ho-on Kai (The Saito Gratitude Foundation) Monographs No. 1. 80 pp.

Mordvilko, A. 1927. Les pemphigiens des pistachiers et leurs forms anocycliques. Mem. Soc. Zool. France 28(315):61–74.

Mound, L. A. and Halsey, S. H. 1978. Whitefly of the world. New York: British Museum of Natural History and J. Wiley & Sons.

Neunzig, H. H. 1964. Notes on the biology of *Eurytoma rhois* complex (Hymenoptera: Eurytomidae). Proc. Ent. Soc. Wash. 66(2):87–88.

Opler, P. 1969. Two new species of *Caloptila* associated with *Rhus* in California (Lepidoptera: Gracilariidae). Pan Pacific Ent. 45(4):249–256.

Patch, E. M. 1938. Food plant catalogue of the aphids of the world, including the Phylloxeridae. Bull. Maine Agric. Exp. Sta. 393:35–431.

Pettey, F. W. 1933. New species of South African Psyllids III. Union S. Afr. Dept. Agric. Ent. Mem. 8:3–23.

Pinhey, E. 1972. Emperor moths of South and South Central Africa. Capetown: C. Struik (Pty) Ltd.

Poos, F. W. and Wheeler, N. H. 1943. Studies on host plants of the leafhoppers of the genus *Empoasca*. U.S. Dept. Agric. Tech. Bull. 393.

Popov, Y. A. 1965. Towards the knowledge of the terrestrial Hemiptera fauna of the southern regions of the western Tien-Shan (USSR, Mid-Asia). Acta Ent. Mus. Nat. Pragae 36:219.

Prentice, R. M. 1962. Forest Lepidoptera of Canada. Vol. 2. Nycteolidae, Noctuidae, Notodontidae, Liparidae. Can. Dept. For., For. Ent. Pathol. Br. Bull. 128:75–281.

Pritchard, A. E. 1953. The Gall Midges of California. Bull. Calif. Insect Surv. 2(2):125–150.

Prokopy, R. 1976. Feeding, mating, and oviposition activities of *Rhagoletis fausta* flies in nature. Ann. Ent. Soc. Amer. 69(5):899–904.

Regas-Williams, K. A. and Habeck, D. H. 1979. Life history of a Poison Ivy Sawfly *Arge humeralis* (Hymenoptera: Argidae). Florida Ent. 62(4):356–363.

Richards, W. R. 1972. *Aulacorthum rhusfoliae* (Homoptera: Aphididae): A new Poison Ivy inhabiting aphid from Ontario. Can. Ent. 105:173–174.

RICHARDS, W. R. 1976. A host index for species of Aphidoidea described during 1935–1969. Can. Ent. 108:499–550.

RILEY, E. 1874. The jumping Sumac Beetle—*Blepharida rhois* (Forst.). Missouri State Entomologist, 6th Annual Report: 118–122.

RIZK, G. N. and Ardini, S. A. 1981. Ecological studies on the pistachio bark beetle, *Chaetoptelius vestitus,* in Iraq. Ain Shams Univ. Fac. Agric. Res. Bull. 0(666):1–11.

ROBERTSON, C. 1896. Flowers and insects. Bot. Gaz. 22:154–16.

ROBERTSON, C. 1929. Flowers and insects. Carlinville, IL.

ROHWER, S. A. 1916. In Viereck, The Hymenoptera, or wasp-like insects, of Connecticut. Conn. St. Geol. Nat. Hist. Surv. Bull. 22.

ROMANENKO, K. E. 1972. The principal pests of fruit of pistachio (*Pistacia*) and the possibilities for their biological or chemical control in Kirgiza. Proc. 13th Int. Congr. Ent. 3:85 (in Russian).

ROMEO, A. 1929. Contributo alla conscenza del "cimiciata" del pistaccio (nota preliminaire). (Rev. Appl. Ent. 22:51 (1934)).

SATTLER, K. 1982. A review of the western Palearctic Gelechiidae (Lepidoptera) associated with *Pistacia, Rhus,* and *Cotinus* (Anacardiaceae). Ent. Gaz. 33:13–32.

SCHAEFER, C. W. 1980. The host plants of the Alydinae, with a note on heterotypic feeding aggregations (Hemiptera: Coreoidea: Alydidae). J. Kans. Ent. Soc. 53:115–122.

SCHERER, G. 1973. Ecological and historic zoogeographic influences on concepts of the genus as demonstrated in certain Chrysomelidae (Coleoptera). Zool. Scripta 2:171–177.

SCHERER, G. 1974. Review of North American species of *Orthaltica* with new generic synonymy (Coleoptera: Chrysomelidae: Alticinae). Coleop. Bull. 28(2):65–72.

SCHNEIDER, F. 1958. Some insect pests of Pistachio in Syria. FAO Plant Bull. 6(5):65–71.

SEVASTOPULO, D. G. 1941. On the food-plants of Indian Agaristidae and Noctuidae (Heterocera). J. Bombay Nat. Hist. Soc. 42(2):421–443.

SMITH-MEYER, M. K. P. 1979. The Tenuipalpidae (Acari) of Africa, with keys to the world fauna. Ent. Mem. Dept. Agric. Tech. Serv., Rep. S. Afr. No. 50.

STEBBINS, F. A. 1910. Insect galls of Springfield, Massachusetts, and vicinity. Springfield Mus. Nat. Hist. Bull. No. 2:1–139.

STEYSKAL, G. 1951. Insects feeding on plants of the *Toxicodendron* section of the genus *Rhus* (Poison Oak, Ivy, or Sumac). Coleopterists Bull. 5(5/6):75–77.

TAKIZAWA, H. 1978. Notes on Taiwanese chrysomelid larvae. V. Ent. Rev. Japan 31(1/2):75–84.

TANG, C. 1976. The Chinese Gallnuts, their multiplication and means for increasing production. Acta Ent. Sinica. 19(3):282–296 (in Chinese with English summary).

TANIGOSHI, L. K. and Nishu-Wong, J. V. 1981. Greenhouse rearing of citrus thrips, *Scirtothrips citri,* for experimental testing. J. Exp. Ent. 74(2):213–214.

TIETZ, H. 1972. An index to the described life histories, early stages and hosts of the macrolepidoptera of the continental United States and Canada. Vol. 2. Sarasota: Allyn Museum of Entomology.

TISSOT, A. N. A new aphid from Poison Ivy (*Rhus radicans* L.). Fla. Ent. 12(1):1–2.

TROTTER, A. 1900. Description de deux nouveaux *Eriophyes* de Chine (Acar.). Bull. Soc. Ent. France 2:179.

TSAI, P. H. and Tang, C. 1946. The classification of the Chinese gall aphids with descriptions of 3 new genera and 6 new species from Meitan Kweichow. Trans. Roy. Ent. Soc. Lond. 97:405–418.

TUCKER, R. W. E. 1926. Some South African mites, mainly Tetranychidae and Eriophyidae. Ent. Mem. Dept. Agric. Pretoria No. 5:1–15.

TZANAKAKIS, M. E. 1962. Some phytophagous Coleoptera of Pistachio. 1. Clytridae, 24 pp. (in Greek); 2. Cryptocephalidae, 17 pp. (in Greek); 3. *Pseudocoeliodes rubricus* (Gyll.) (Coleoptera: Curculionidae), 20 pp. (in Greek). Rev. Appl. Ent., Ser. A., 1964, 52:238 (English Abstract).

ULU, O., Zumreoglu, O., and San, S. 1974. Preliminary studies of the pests of Pistachio nuts and their parasites and predators in the Aegean Region. Plant Prot. Res. Ann. 6:55,184 (in Turkish). English summary in Rev. Appl. Ent., Ser. A, 1974, 64(10):1188 (no. 4480).

VAN SOMEREN, V. G. L. 1974. List of food plants of some east African Rhopalocera, with notes on the early stages of some Lycaenidae. J. Lepid. Soc. 28:315–331.

VAN SOMEREN, V. G. L. and Rogers, K. S. 1935. The butterflies of Kenya and Uganda (continuation and conclusion). J. E. Afr. Uganda Nat. Hist. Soc. 12(3/4):59–89.

Van Son, G. 1979. Butterflies of Southern Africa, Part 4, Nymphalidae: Nymphalinae. Transvaal Mus. Mem. No. 22.

Vari, L. 1961. Southern Africa Lepidoptera. Vol. 1, Lithocolletidae. Amsterdam: Swets and Zeitlinger.

Walt, J. F. and McPherson, J. E. 1973. *Stethaulax marmoratus* (Say) (Scutellaridae) on *Rhus glabra* and *R. copallina* drupes in Illinois. Ann. Ent. Soc. Amer. 66(5):1103–1107.

Weber, B. C. and McPherson, J. E. 1983. World list of host plants of *Xylosandrus germanus* (Blandford) (Coleoptera: Scolytidae). Coleop. Bull. 37(2):114–134.

Wertheim, G. 1954. Studies on the biology and ecology of gall-producing aphids of the tribe Fordini (Homoptera: Aphidoidea) in Israel. Trans. Roy. Ent. Soc. 105(5):79–96.

White, B. E. 1942. A new genus and species of Coleoptera (Chrysomelidae) from the southwestern United States. Ent. News 53(1):16–21.

Wilcox, J. A. 1979. Leaf Beetle host plants in northeastern North America. The Biological Research Institute of America, Inc. Kinderhook (New York), World Natural History Publications.

Williams, M. L. and Kostarab, M. 1972. The Insects of Virginia: No. 5. Morphology and systematics of the Coccidae of Virginia with notes on their biology (Homoptera: Coccidae). V. P. I. S. U., Res. Div. Bull. 74.

Wiltshire, F. P. 1957. The Lepidoptera of Iraq. London: Nichalas Kaye Ltd.

Wood, S. L. 1982. The Bark and Ambrosia Beetles of North and Central America (Coleoptera: Scolytidae), a taxonomic monograph. Great Basin Nat. Mem. No. 6.

Worth, C. Brooke, Platt, A. D., and Williams, F. F. 1979. Differential growth among larvae of *Citheronia regalis* (Saturniidae) on three genera of foodplants. J. Lep. Soc. 33(3):162–166.

Xiang, H. 1980. Chinese Gallnut aphids on *Rhus potaninii*. Entomotaxonomia 2(4):303–313.

Young, D. A. 1972. Reproductive biology of *Rhus integrifolia* and *R. ovata*. Evol. 26(3):406–414.

FIGURES .

Fig. 1. Individual tree of *R. tripartita* ca. 3 m in height in Judean Desert, Wadi Mukallik, 18 January.

Fig. 2. Typical scrubbed form of Judean Desert *R. tripartita,* ca. 0.5 m in height (insect net = 15 inches diameter), Wadi Qilt, 11 January.

Fig. 3. Goat grazing causing dwarfed appearance of Sumac trees in Judean Desert, W. Qilt, 11 November.

Fig. 4. Typical habitat of relictual desert populations of *R. tripartita* (= few isolated trees on canyon slopes), W. Qilt, 14 April.

Fig. 5. Typical habitat of refugia Mediterranean populations of *R. tripartita* (= large trees in continuous hillside populations), Sanniriya, Western Samaria Mountains, 4 April.

Fig. 6. Typical robust form of Mediterranean *R. tripartita* (two trees more than 2 m high at Yarhiv in Samarian foothills overlooking Mediterranean coastal plain), 4 April.

Fig. 7. Crawler of the mealybug *Peliococcus deserticola* feeding on the roots of *R. tripartita* at Wadi Qilt, 18 December.

Fig. 8. Scales (primarily females), *Melanaspis inopinatus* on branch of *R. tripartita* at Wadi Qilt, 18 December.

Fig. 9. Parasitic plant (*Loranthus acaciae*) on branches of *R. tripartita* at Wadi Qilt, 11 January.

Fig. 10. Larva of the monophagous leaf beetle *Blepharida sacra* (note "fecal thread"), from Yarhiv, 1 January.

Fig. 11. Cluster of mid-instar larvae of *Thaumetopoea jordana* on *R. tripartita* leaves at Wadi Qilt, 18 January.

Fig. 12. *R. tripartita* leaves rolled for feeding by larvae of *Teliopsis sarcochroma* (left of center = older discolored leaf-roll, right of center = freshly rolled leaf) at Wadi Qilt, 14 April.

Fig. 13. *R. tripartita* leaf galls caused by the eriophyid mite *Aculus* ? sp., at Yarhiv, 28 October.

Fig. 14. *R. tripartita* fruits infested by a torymid wasp (*Megastigmus pistaciae*), at Yarhiv (the few lighter colored fruits are normal un-infested, most fruits infested and are black or gray in color), 4 April.

Fig. 15. Cryptically colored crab spider (*Thomisus ornatus*) waiting for prey on cluster of male *R. tripartita* flowers at Yarhiv, 8 November.

Fig. 16. Lizard (*Eremias guttulata*) impaled on Sumac thorn as prey by the Great Grey Shrike, W. Qilt, 18 January.